T0092790

Artificial Intelligence and The Environmental Crisis

Can Technology Really Save the World?

Keith R Skene
Director, Biosphere Research Institute
Angus, Scotland, UK

A SCIENCE PUBLISHERS BOOK

CRC Press
Taylor & Francis Group
6000 Broken Sound Parkway NW, Suite 300
Boca Raton, FL 33487-2742

© 2020 by Taylor & Francis Group, LLC
CRC Press is an imprint of Taylor & Francis Group, an Informa business

No claim to original U.S. Government works

Version Date: 20191024

International Standard Book Number-13: 978-0-367-15200-0 (Hardback)

This book contains information obtained from authentic and highly regarded sources. Reasonable efforts have been made to publish reliable data and information, but the author and publisher cannot assume responsibility for the validity of all materials or the consequences of their use. The authors and publishers have attempted to trace the copyright holders of all material reproduced in this publication and apologize to copyright holders if permission to publish in this form has not been obtained. If any copyright material has not been acknowledged please write and let us know so we may rectify in any future reprint.

Except as permitted under U.S. Copyright Law, no part of this book may be reprinted, reproduced, transmitted, or utilized in any form by any electronic, mechanical, or other means, now known or hereafter invented, including photocopying, microfilming, and recording, or in any information storage or retrieval system, without written permission from the publishers.

For permission to photocopy or use material electronically from this work, please access www.copyright.com (http://www.copyright.com/) or contact the Copyright Clearance Center, Inc. (CCC), 222 Rosewood Drive, Danvers, MA 01923, 978-750-8400. CCC is a not-for-profit organization that provides licenses and registration for a variety of users. For organizations that have been granted a photocopy license by the CCC, a separate system of payment has been arranged.

Trademark Notice: Product or corporate names may be trademarks or registered trademarks, and are used only for identification and explanation without intent to infringe.

Visit the Taylor & Francis Web site at
http://www.taylorandfrancis.com

and the CRC Press Web site at
http://www.crcpress.com

Dedication

This book is dedicated to two inspiring, precious friends whose passing has left such a void in the lives of all of us who loved them

Klement Rejšek

Mark 'Nebbs' Turnbull

Preface

"I went to the crossroad,
fell down on my knees
I went to the crossroad,
fell down on my knees
Asked the Lord above
have mercy, now save poor Bob, if you please"

Cross Roads Blues, Robert Johnson, 1936

We find ourselves at one of the great intersections of our short but eventful history. One road, the path of incessant greed and selfishness, meets the other, the path of our increasingly ailing planet. It is a collision course, and we are unlikely to survive the impact intact. Our highway of consumption has laid waste to much of the Earth as we have accelerated like there was no tomorrow, using new technologies to optimize conditions for ourselves, while maximizing profit from draining the resources of our world.

We have cut ourselves off from nature, and have ignored the warnings of a silent spring, lured by the songs of the Sirens of pleasure, luxury and apathy. The decision facing us is whether we ignore the devastation and plough steadily onwards and downwards, or whether we turn away from this ruinous path. Sustainable thinking is no longer a sweet folk song, replete with Appalachian dulcimers and autoharps, a sing-a-long chorus and less familiar verse, interchanging on a theme of the beautiful landscapes of our youth, infused with a Gaian spirit. Sustainable thinking is essential thinking, if we are to avoid the calamitous crash that lies ahead. Our current path is unsustainable, and the threats are multiple.

In this book we directly address the thorny issue of whether all technology is fundamentally bad for the planet, or if, in fact, technology could be the life raft in a tumultuous sea of environmental crisis. Throughout our journey, technology has undoubtedly helped deliver the cataclysmic collapse of the Earth system on which we depend, helping us to enslave the very soul of our world for our own wasteful gratification. We have used this technology to damage, almost beyond repair, the exquisite but delicate synchrony of the Earth system. Yet it is not the technology that instigated this devastation, but ourselves. This book is a call to arms, challenging us to take responsibility for our predicament and recognizing that the Earth system is the only ark in town. Do we join it or do we exclude ourselves from a future within it?

At the forefront of technology today lies artificial intelligence. The field is over sixty years old, but recent progress has transformed it into something we all need to know about. We are now living in the infosphere, that parallel digital universe that is becoming more and more a part of the real world. Here, information is king. We are surrounded by seven billion devices that, together, make up the internet of things. These appliances are silently gathering data on us continuously and feeding it into huge data repositories. Everything from smart electricity meters to satellites orbiting the Earth convert our activities into binary code, shepherding this vast unruly herd of data into arrays, ready for analysis.

The infosphere feeds off this cornucopian feast of big data and is dominated by a few powerful companies worldwide. It has the potential to do immeasurable good, and to fundamentally restructure our very evolution on the planet. There is also increasing concern that it could deliver a modern-day apocalypse, posing an existential threat.

The thing about technology is that it does what we tell it to do, and within the limits that we set it. Or at least it does at present. But what of the future? What if artificial intelligence really did become autonomous, sentient and creative? By making its own decisions it could terraform our planet into a wholly different reality. Some fear that this brave new world might not have a place for us in it at all. We could be viewed as liabilities. What if the infosphere was no longer a mirror world, a parallel representation in silicon and binary, but rather that it became as flesh and dwelt among us, or displaced us?

This book asks some difficult questions. The questions don't really focus on artificial intelligence, but on ourselves. If we base this potentially transformative technology on the intelligence of humans, then how can we expect it to deliver a sustainable transition that is so urgently needed, given the flawed nature of our own thinking? The planet is shrouded in inequality at every level, be it power, wealth, opportunity or health. Not only are we destroying the environment, but also society. Are we really the best model for such a potentially potent technology?

Could there be a different model of intelligence on which to construct our futures? How about an intelligence that has been around for over three billion years, honed from the toughest of life experiences, an intelligence that has recovered from ice ages and mass extinctions. An intelligence that has directed the recovery from almost complete wipe outs, such as the snowball Earth and that catastrophic comet that finished off the dinosaurs?

This intelligence is composed of multiple different intelligences, gathered together in an interactive whole, operating as one, and continuously learning, evolving, diversifying and sharing. It is a superintelligence that is resilient and resistant, and that holds within it the ultimate wisdom of physics, with knowledge as old as the Universe itself. Not a religion, nor a cult or sect, but the lifeforce that flows through the biosphere: ecosystem intelligence.

What if we rebooted AI to resonate and flow with the intelligence of the Earth? This book sets out this vision, wherein a properly grounded AI can be the catalyst to re-connect us to our planet, rather than further ostracise us from it. The Earth already has a perfectly good intelligence network, tried and tested over eons. It works, it's free and it's brilliant. There's no need to re-think things. All we need to do is to understand it and plug back into it. Re-integration, not re-invention, is the way ahead.

There are reasons why we haven't recognized this, and we will deal with these in order to identify the real barriers to transition. As Shakespeare once wrote, "*The fault dear Brutus is not in our stars, but in ourselves*". So, we'll search deep within ourselves to find how to break free of our constraints.

AI can be the perfect technology to deliver a new age of sustainability, but not as it stands at present. By rebooting AI, basing it on ecosystem intelligence and plugging it into the biosphere physically and metaphysically, this book argues that we can move into a new era, where re-integration can deliver a sustainable future. We find ourselves at the crossroads. This is the time and the place for change in our journey, a point of no return. Continued failure is not an option. We need to make the right call, or like the great bluesman, Robert Johnson, we may have to sing the final refrain:

> *"That I got the crossroad blues this mornin',*
> *Lord, babe, I'm sinkin' down."*

Keith R Skene

Contents

SECTION I Artificial Intelligence and the Internet of Things

"The most profound technologies are those that disappear. They weave themselves into the fabric of everyday life until they are indistinguishable from it."

–Mark Weiser, *Scientific American* (Weiser, 1991)

Welcome! Diversity is a theme that will run through this entire book, as we explore the importance of this key concept in society, ecology and technology. Diversity lies at the heart of resilience and creativity, broadening the available solution space. Indeed, our ability to solve problems is more likely to be limited by a lack of diversity in thinking than by limitations imposed by the problems themselves. Perhaps the greatest challenge for computer technologists is how to embed and maintain diversity into artificial intelligence (AI), in a globalized, connected infosphere. The infosphere is the new technological world, where information, in digital form, creates another dimension of reality. This virtual space is increasingly affecting the 'real' world, and the two are beginning to merge.

All of us interact with the world around us in different ways as an outcome of our genes, our journey and our landscape. We also have different levels of knowledge on the subject matter of this book. Some will be very aware of artificial intelligence, sustainability, ecology and social science, others less so. This section is designed to bring everyone to a similar point at the outset. If you are already comfortable with areas such as symbolic and non-symbolic approaches, the internet of things and the issues surrounding whether intelligence can be artificial at all, then feel free to skip this section. Otherwise, read on.

In the next few pages, you will travel back in time to the earliest conception of artificial intelligence, some three thousand years ago, in China and Greece, before relocating to a soda machine in a Brutalist university building in Pittsburgh, the birthplace of the internet of things. Then we take a whistle top tour of the terminology surrounding artificial intelligence, while at the same time following its development. We'll meet some of the key characters along the way, from King Mu of Zhou to Marvin Minsky. We trace the euphoric rise, the barren AI winter and the resurgence of a field that most of us have only really viewed through the lens of a Hollywood blockbuster movie camera. So, let's make a start.

1.1. Nothing new under the Sun

We think of artificial intelligence as a really modern phenomenon, at the cutting edge of technology, representing the dawn of a brave new world, or an apocalyptic vision of robotic dominion and terror. Yet the concept of artificial intelligence actually reaches much further back in time, some three thousand years before present. Almost as soon as written records started appearing, humans have envisioned machines that could think, feel and act just like we can. Two separate strands of evidence show that, early in recorded history, people could imagine artificial life forms. The first of these strands comes from Ancient China, while the second emerged from Ancient Greece.

The *Liezi*, sometimes referred to as the *True Scripture of Excellence*, is a revered Daoist scripture and philosophical treatise, viewed as one of the three most important texts in Taoism, along with the Laozi and the Zhuangzi. Written around two thousand five hundred years ago, and recomposed one thousand years later, it consisted of around one hundred and forty parables. The Liezi is accredited to Li Yukou, the Daoist philosopher. Daoism emphasises the importance of achieving a resonance between our inner being and the outer, natural world, leading to a holistic integration.

While some question if Li Yukou ever actually existed, and others question if he wrote the text, or if it was merely a collection of plagiarised passages from other classic texts, the fact is that the book itself exists and is of great interest at many levels. However, in relation to our purposes, it is a story revolving around the wonderfully-named King Mu of Zhou that is of interest.

King Mu was a real king who ruled China around 950 BC. In Book Five of the Liezi, the story is told of the king meeting an engineer called Yen Shih, who had built a human-like machine that could walk, sing and dance. The king was really impressed, until the robot made advances at his concubines and winked and cavorted with them. King Mu ordered the execution of the engineer, and he was only saved when he destroyed the robot. We note in passing that this early account of an intelligent machine was laced with lust and jealousy. It is also of interest that King Mu did not hold the automaton responsible for its desire and depravity, but rather the engineer, who was sentenced to death, briefly. This will become of relevance when we consider ethical accountability, later in this book.

The second early conception of artificial intelligence occurs in the work of Homer. Questions exist over whether or not Homer even existed as well as whether or not he wrote the *Iliad* and *Odyssey*, the two great classic works of Ancient Greece. However, we do know that Troy existed, and that the Iliad was written in the 8th century BC. The dating of this work is as interesting as much of the content. In a similar way that genetic mutations through time have been used to elucidate when species separated from other species, so linguistic experts have been able to trace changes in language over time, i.e.

linguistic mutations, and determine when particular works of literature were written. According to Eric Altschuler and colleagues, the Iliad was written sometime between 710–760 BC (Altschuler et al., 2013). It is considered to be the first work of the Western literary tradition.

The Iliad focuses on events surrounding the fall of Troy, an ancient city in what is now North West Turkey. The fall of Troy itself is generally thought to have occurred in the twelfth century BC. Of particular interest to us is an account of the banqueting hall of the gods, and a series of handmaids made out of gold. Mounted on tripods with gold wheels, they served the tables of the gods at a great feast. They were made by Hephaestus, the blacksmith god who also created Achilles' armour. These structures were more than mere golden statues. Not only could they move around and serve food independently, but they had the ability to speak. Homer states that they had *"understanding in their hearts"* (Iliad XVIII, 417-425). These were machines with understanding in their hearts. The concept of an advanced artificial intelligence had been conceived of in Ancient Greece as well as in China, almost three thousand years ago.

As Pablo Picasso, the Cubist painter, is reported to have exclaimed (though this is disputed), when viewing the ancient cave art in Lascaux in 1940, with its clever use of twisted perspective, for which Picasso was famous, *"We have invented nothing"* (Graff, 2006). As a noteworthy aside, the caves at Lascaux were discovered by some young local schoolchildren when their dog, Robot (indeed!) disappeared down a hole near the village they lived in. The children followed, in search of the dog, and found themselves in a suite of caverns, surrounded by vast, painted panoramas, unseen for thousands of years.

In the *Odyssey*, the second of Homer's epic poems, that describes the return of the conquering heroes from Troy, another example of artificial intelligence is found, but without the issues of lust or jealousy. Homer writes of the boats of the Phaeacians, a hedonistic people from the modern-day island of Corfu: *"Phaeacian ships have no helmsman or steering oar, for the ships themselves know our thoughts and wishes, and the cities of men, every fertile country, and hidden by mist and cloud they speed over the sea's wide gulf, and never fear damage or shipwreck."* Not merely a satnav or autopilot here, but a boat that knows our thoughts and wishes.

And so, the dreams and conception of artificial intelligence have been with us for millennia and straddle the cultural globe. The Syrian writer, Lucien, was the author of what is considered the first sci fi novel, *True History* (Hickes, 1894). Lucian lived during the second century AD, and hailed from Samosata, a city whose remains now lie under the waters of the Atatürk Dam in southeast Turkey.

True History is a satirical classic, reminiscent of the eighteenth-century French tale, *Candide, oul'Optimisme*, whose author, François-Marie Arouet, was better known by his nom-de-plume, Voltaire. Both books ridicule supernatural religious practices. The work of Dante Alighieri, Jonathan Swift,

Ludovico Ariosto, Cyrano de Bergerac, Edgar Alan Poe, Jules Verne and Herbert George Wells all owe much to Lucian's ground-breaking approach.

In *True History*, at one point, Lucian and his crew arrive at Lychnopolis, or lamp city, apparently located in the constellation of Taurus (the bull) between two different star clusters, the Hyades (at the base of the horns of Taurus) and the Pleiades (at the tail of Taurus). Here they encounter a series of highly knowledgeable lamps. Finding one that represented Earth, Lucian writes that they "*spake unto it and questioned it of our affairs at home, and how all did there, which related everything unto us*".

And so, we see that the concept of artificial intelligence has been around for a very long time. But it was not until the twentieth century that these visions could be made in flesh and blood, or, rather, silicon and gold.

We now get to grips with some of the jargon associated with artificial intelligence, the context and development of the field and, more fundamentally, we define what is meant by artificial intelligence. If you are already comfortable with these areas, you can probably skip or skim this section. However, if you are only starting to study the whole subject, then it is probably essential reading. I'll try to make it as pain-free as possible.

First, some definitions. There is a glossary at the back of the book that you can access at any time, but let's deal with some of the main players here. When thinking about artificial intelligence, we are really dealing with two complementary strands, the *internet of things* (IoT) and *artificial intelligence* (AI). So, let's get started.

1.2. Oh, for a nice cold soda: The birth of the internet of things

Wean Hall is a big box of a building of rough-hewn concrete, an example of the Brutalist architectural tradition. It sits in a Beaux-Arts quadrangle, as if dropped from a height by modernist aliens just to offend and provoke. So ugly is it that the architect is unknown. I guess no-one wants to be associated with it (unless it was actually dropped by aliens). It houses the Department of Computer Science of Carnegie Mellon University (CMU) in Pittsburgh. Famous for being the alma mater of Andy Warhol, CMU has an even greater claim to fame: the home of the first 'thing' in the internet of things (or one of the first anyhow).

Computing science is thirsty, warm, long work. The hardware releases a lot of heat. It's also compulsive, and hours go by. Caffeine and cold beverages are essential. And in Wean Hall, there was a soda machine, promising both of these things in one bottle. I remember cramming for exams at the University of Illinois, Urbana-Champaign, on a diet of Mountain Dew and Pro Plus. But in Wean Hall, the soda machine was on the third floor, while the research labs were on the fourth floor. Stuck at your work station for hours, you become aware that your mouth is as dry as a frog in the Sonoran Desert. You crawl, heavy-limbed and weary, downstairs to the machine of your dreams.

You know it's going to be OK. A light-headed sensation of hedonistic anticipation begins to build, elation growing as if you'd crawled across the aforementioned desert and were just one sand dune away from the oasis of heavenly delights with its pool of cold, effervescent liquid. Refreshment and awakening lay just along the corridor, down the stairs and a few short steps to the left. But imagine the overwhelming sorrow if, when you got there, armed with your nickel, the machine was empty. It's enough to fell a human being to the ground. An old line from an ancient hymn comes floating through the air.

"I tried the broken cisterns, Lord,
But, ah, the waters failed!
E'en as I stooped to drink they fled,
And mocked me as I wailed."

Or it could be even worse, if that is possible. Just before you descended the stairs to the corridor of bounteous beverages, one of the student volunteers had filled the machine with bottles from the hot store cupboard down the corridor. You arrive. It looks perfect, just how you imagined it would. Bottles all lined up bolt-upright, ready to satiate your cravings. A scene of true glamour and perfection, like a painting by the great former student, Warhol, himself. You deposit the nickel, reach down for the bottle and your world darkens, as thermoreceptors in the skin on your fingers transmit the devastating news to your brain by way of a series of electrochemical impulses, that it interprets and whispers to your conscious self: The.bottle.is.warm.

There is nothing worse than a warm soda. There's no way around it. It will take three hours to chill. But you've already removed it from the chiller, and that was your last nickel. You stand in the corridor, forlorn and friendless, and memories of all of the lowest, emptiest, most isolated feelings from your life and all the other crumpled lives that ever slumped against this wall through time immemorial come tumbling down around you.

You realize that you've slid to the floor, but you don't care. You groan that sound that comes from some deep cavern within your soul, but far more extensive than the known universe and the universes beyond. This sound is not the Vedic *'Om'*, proclaiming the creative powers of the universe, but the polar opposite. This is not the big bang, but the thirsty, overheated, very, very sleepy big shrink. Dreams dashed, devastated and unsatiated, you feel exiled, brutalized and so very alone.

Yet, in our lowest moments, hope is the last to die, and with hope, springs imagination. In 1971, high in the Brutalist box that is Wean Hall, a computer scientist had a dream: to know if the soda machine had soda in it and to know how cold that soda was, without even leaving his work station. David Nichols, a graduate student, convinced two other students, Mike Kazar and Ivor Durham, as well as a research engineer, John Zsarnay, that they had the technology to make this dream a reality.

The soda machine, like all the other soda machines in the world, had indicator lights. When a drink was bought, a red light would flash on and off. If the last bottle on the rail was purchased, the light would stay on. Once the drinks had been replaced, the light would go off again. The students installed electronic sensors to monitor the activity of the lights in the machine, and a cable ran from the sensors to a mainframe computer. From here, data could be fed through the forerunner of the internet, called the Advanced Research Projects Agency Network (ARPANET).

The ARPANET linked around three hundred computers, mostly in the USA. As a result, anyone, from Santa Barbara, California to Cambridge, Massachusetts, could check on the status of the soda machine in Wean Hall. But, most importantly, the computer scientists upstairs could know if there was cold soda down below. The first really useful *smart device* was born. A smart device is an electronic device that can connect, share and interact with its user and other smart devices through the internet. A revolution had begun.

Today, there are over seven billion smart devices in the world, not including smartphones, and the numbers continue to soar. The British Government has a target to have smart electricity meters in every home by 2020. These devices are continuously feeding data into the internet. They are the eyes and ears of the infosphere, listening and watching, measuring and mapping our every move. Connected together, they form what is called the internet of things, or IoT, a term coined, in 1999, by the executive director of Auto-ID Laboratories, Kevin Auston.

Other names have been used, such as the web of things, internet of objects, embedded intelligence, cyber physical systems, pervasive computing, ubiquitous computing, calm technology, machine-to-machine (M2M) and human-computer interaction. But it is the internet of things that has stuck. Defined as *"an open and comprehensive network of intelligent objects that have the capacity to auto-organize, share information, data and resources, reacting and acting in face of situations and changes in the environment"* (Madakam, 2015), the IoT has become synonymous with the digital age. It is now embedded into many elements of critical infrastructure. Increasingly invisible, it cohabits the world with us, often without us even knowing. We live in smart houses, in smart cities, working in smart offices and watching smart televisions. Soon we'll all be commuting in smart cars and hanging out with smart robotic companions.

This pervasive connectivity, linking automation, integration and servitization, is steadily growing, to the extent that the US National Intelligence Council included the IoT as one of its six disruptive civil technologies, the other five being *clean coal technology* (posing a threat to the oil-dependent economy of the USA), *robotic technology* (threatening jobs and social fabric), *biofuels* (requiring dependence on other countries), *energy storage technology* (again threatening the oil-based economy) and *biogerontechnology* (increased life expectancy through technology, having huge cost implications and demographic consequences).

IoT has perhaps been the most sensitive area, particularly militarily, as it represents potential information security vulnerability. Information is the new oil, and represents power and control. This is why organizations such as Wikileaks are viewed as significant threats. Companies that control the connectivity of the internet are seen as particularly powerful, as they control the internet of things. The new 5G mobile internet connectivity is creating concerns among cyber-security experts. The potential problems arise because the companies that control the connectivity have the potential to tap into what is being communicated. The eyes and ears of the internet could become the eyes and ears of these companies, realizing information security risks. Even more concerning to national governments is the situation where such companies are under the control of potentially hostile governments. However, the cloak-and-dagger, allegation and counter-allegation world of industrial and military activity makes it difficult to know exactly where we sit.

It's a bit like the *Potemkin village*, as my precious, sadly departed friend, Klement Rejšek (to whom this book is dedicated), explained to me, while discussing Russian politics. Catherine the Great was Russia's longest reigning female monarch, ruling from 1762 to 1796. Her reign coincided with the Golden Age of Russia. Famous for her brilliant political strategy, she often had affairs with her most capable ministers.

One of the most important supporters of her reign was Grigory Potemkin, the military leader and nobleman. By 1787, he was governor of Crimea. Crimea had recently been taken from the Ottoman Empire after a bloody war. Potemkin wanted to attract Russians into the newly acquired territory in order to build stability and trade.

The story goes that Catherine the Great was due to visit in order to help her former lover with his mission. Rather than Catherine seeing the widespread starvation, poverty and ruin, Potemkin arranged for a mobile village to be assembled, with fake walls, fake villagers (his well-fed troops in disguise) and fake food on display. Catherine toured by boat, not landing at the village, and so could not see that it was fake. Each night, Potemkin had his troops disassemble the village and move it downstream, where, the following day, Catherine would again encounter what she thought was another perfect village.

The Potemkin village has been used as a metaphor in economics and politics, representing a subterfuge that appears to show things being better than they are. Of course, if it suits, the opposite trick can be as powerful, pretending things are bad or questionable in order to create distrust in a company, political party or country. And what an Achille's heel the connectivity of the internet of things represents: the very oxygen of the infosphere.

Yet spreading a rumour that oxygen is poisonous may not be the cleverest way to proceed. The few companies that can actually supply 5G are already heavily involved in many aspects of mobile technology service provision, to such an extent that it would be difficult to remove them, potentially disrupting this precious connectivity. Indeed, the internet of things is fundamental to any global business today, collecting data that allows predictability and foresight and to follow objects through the commodity chains in which they are situated. As we have already noted, information is power. The transfer of information is the life blood of our modern world.

And here lies the great irony. While companies want to gain information from competitors and consumers, they don't want their own information to be shared with anyone else. Unshared information is another form of power. But it is difficult to have it both ways. You can't have both off-grid privacy and interconnected sharing. Other issues with IoT include diverse standards and different technologies across the globe, leading to what are called *fragmented ecosystems*, wherein true functional integrity is not possible. Also, the sheer rate of expansion and modernization creates its own difficulties, as docontinuous, though necessary, regulatory changes.

Secure connectivity is crucial in order to build confidence and trust amongst increasingly connected humans, to avoid data theft, to protect health and safety, to prevent a loss of productivity and to avoid noncompliance to regulations targeting the aforementioned areas. Another challenge is that as the IoT expands, there is the increasing likelihood that devices will be connected to the system that have been set up by inexperienced coders, insufficiently schooled in the fine arts of cybersecurity. This can allow an entrance point for attackers, a weak link in the chainmail of the IoT knight. As the fearsome biblical warrior Goliath found, even a small shepherd boy armed with a sling and a pebble can land a killer blow if he hits a weak spot.

Many of the devices now connected are battery-powered and cheap, with insufficient power to host adequate security programmes. Also, it is increasingly recognized that software alone cannot provide the protection needed to ensure security. More often now, hardware security circuitry or parallel security processors are in use, but this again increases costs and power requirements.

As the IoT becomes a key player in essential and critical infrastructure, its value increases in terms of criminal interest. Vulnerability means leverage in terms of malevolent intentions, where the ability to disrupt technology becomes extremely lucrative. Technological hostage situations are already occurring, where individuals or companies are asked to pay a ransom, often in Bitcoins, in order to have services restored.

So, the IoT, born in the minds of four thirsty, weary computer scientists in Pittsburgh wanting a cold soda, has become a global Argos, the many-eyed giant of Ancient Greece, whose eyes never close (even without a caffeinated soda).

The concept is expanding. The internet of everything (IoE) refers to the interactions between the 'thing', people, data and process. Some argue that this is really just IoT, but others feel that it is more inclusive and allows sociological contexts to be discussed as part of the landscape (Evans, 2012). Smart cities are viewed as large scale expressions of the IoE (Mitchell et al., 2013). Interestingly though, there is no room for the biological environment here.

But it is a second great technological development of the infosphere that promises to truly transform our futures, for better or for worse. That development is *artificial intelligence*.

I.3. The two-month, ten-man project to transform the world

On first appearances, the meaning of the term *artificial intelligence* would seem to be self-evident: a form of intelligence that is artificial, or non-organic. The term was invented by John McCarthy and colleagues, in 1955, in a proposal for a conference (what would become the first Dartmouth Conference) where they wrote: "*We propose that a 2 month, 10 man study of artificial intelligence be carried out during the summer of 1956 at Dartmouth College in Hanover, New Hampshire. The study is to proceed on the basis of the conjecture that every aspect of learning or any other feature of intelligence can in principle be so precisely described that a machine can be made to simulate it*" (McCarthy et al., 1955). They were granted the funding and the historic meeting went ahead. Two months and ten men. It is reminiscent of the two small fishes and five loaves that fed five thousand. From such humble beginnings, a future world was conceived.

Yet, immediately, we are faced with some questions. What do we mean by intelligence, and can intelligence be artificial at all? Surely you are either truly intelligent, or not? Is artificial intelligence some pretence? Is it like an artificial limb, a dental implant, a lawn made of plastic, a bunch of flowers made from silk stretched over metal wire? Artificial flowers may look very much like the real thing, but they cannot be pollinated, do not produce nectar and will never produce seed. So, is artificial intelligence some sort of mimic, some counterfeit concept, unable to deliver the heavy fruit that only hangs from the bough of the true tree of wisdom?

Also, if it is a mimic, what kind of intelligence should AI be seeking to mimic? Is intelligence a strictly organic property, existing only in living organisms? As organisms increase in complexity, from viruses to bacteria, protists, fungi, plants and animals, at what point does intelligence occur and does it differ at different levels of organization? Can a unicellular organism really be intelligent, or do you need to be multicellular? Vegetarians think that animals are more 'special' than plants and should not be eaten, because they feel pain and are sentient. Yet a plant is an extremely complex and, many would say, intelligent sort of organism. Do you have to have a brain, or at least a proto-brain, or can plants be intelligent?

Was there a moment in the evolution of life on Earth when the first intelligent being suddenly popped up, and proclaimed *"you guys, you just don't get it, do you? 'Cause you just don't have it up here, right"*? If not, then either everything is intelligent or nothing is intelligent.

The same goes for species. Imagine lining up all of your ancestors in a row. First would be your mother and father, then, behind them, your grandparents. Next the great-grandparents and so on. You then step out and start walking along the line that stretches far into the distance. I've often thought about this. As you walk along, the fashions would change over time. Eventually, your distant relatives would be dressed in animal furs, or, further back, nothing at all (an awkward moment). Houses would become more primitive, until they were living in caves. The landscape might well change as you discover that many of your ancestors came from other countries or continents. Further back still, you start feeling the chill of an ice age, or the thirst of one of the great droughts that destroyed the Mayan civilization, the Mesopotamian Akkadians or the Egyptian Old Kingdom.

At what point in this walk, as you feel more and more distant from the individuals you are encountering, would one of your relatives try to eat you? At what point would you decide that they were no longer the same species as you? Putting it another way, did a non-human give birth to a human at some point, or a non-lion to a lion? When does a species become a species? This is very hard to grasp. Yet if we tie intelligence to being human, then at that very moment, when the first human was born, and realized its parents were a different species, we would have to say that at that moment too, human intelligence was also born.

These are difficult issues. Yet they are part of a whole set of similar questions. Biology deems that viruses are not living, but that bacteria are alive. To be in the club of life, science deems that you have to pass a number of tests. These are, in no particular order, the ability to reproduce, responsiveness to the environment, homeostasis, trait inheritance, metabolism, development and cellularity. A mule is the offspring of a male donkey and a female horse. Yet a mule is not able to reproduce. If it stares at you, or kicks you, it seems very much alive. However, according to our definition of life, it isn't alive because it cannot reproduce. Yet no one would say that mules aren't alive. Putting this another way, if you kill a mule, have you really killed at all? It seems obvious that you have killed the mule but, strictly speaking, this definition of life might say otherwise.

Viruses are tripped up on their way into the club of life by the cellularity issue in particular. Yet viruses do very well for themselves, and have outwitted us on many occasions. They use our cells, so don't need to be cellular. The Spanish Flu killed more people at the end of the First World War than all of the conflict of the previous four years. Viruses have been around for hundreds of millions of years, surviving mass extinctions and the rise and fall of countless species of cellular organisms. Surely to claim they are not alive is a bit churlish?

I suspect that the real reason behind this is that we don't like to have anything too different from us in the club. At least bacteria are cellular, just like us. If we include viruses, then we might have to include prions. Prions are short strands of protein that cause diseases such as bovine spongiform encephalopathy (mad cow disease). Then we really would have trouble, because proteins are just molecules, and so where would it stop?

Some turn to *panspermia*, the theory that life came from another planet. It kicks the can down the road a bit more. But at some point, we have to face up to it. At some point, life started, and if it didn't, then is there such a thing as life at all? We like to think we know things, but the beginnings are much more difficult that we realize. We conceptualize the Big Bang as the start of everything, but what happened before that? Even with the big bang, big shrink model, wherein the universe has continuously expanded (the big bang phase) until it cannot support itself and has collapsed (the big shrink), which came first, and what came before that? This is cosmology at its most bewildering.

If life emerged on the planet from biochemical events, at what moment was the first living organism actually born, and how? We don't like continua, because we like a beginning and an end. We like cause and effect, but imagining how the first cause came into existence to have the first effect is far beyond us. And it's the same with intelligence. What was the first intelligent organism and when did it arise?

A fairly typical definition of AI is as follows: "*Artificial intelligence (AI) is a variety of human intelligent behaviors, such as perception, memory, emotion, judgment, reasoning, proof, recognition, understanding, communication, design, thinking, learning, forgetting, creating, and so on, which can be realized artificially by machine, system, or network*" (Li and Du, 2017). Mikolov et al. (2016) set out four basic requirements for AI:

1. Interactive communication using natural language;
2. Channelling of non-linguistic information such as sensory perception;
3. The capacity to learn;
4. The capacity for motivation.

Driving the rapid progress towards these goals are the huge amounts of data (*big data*) streaming in from the internet of things and cheaper processing and storage facilities.

We'll examine what it means to be intelligent, and what kinds of intelligence exist in Section III, when we question the choice of human intelligence as the foundation of AI. But let's examine AI in a bit more detail first, because within it lie a number of definitional complexities that we need to be clear on.

I.4. Getting to grips with the jargon: Symbolic and non-symbolic AI

Fundamentally there are two forms of AI: the autonomous, thinking, feeling, sentient genius that is still the stuff of science fiction, and the more mundane, but still very impressive current models like IBM's chess-playing wizard, *Big Blue*, and IBM's quiz mastermind, *Watson*. These two categories of AI are extremely different beasts.

The first type, the autonomous thinkers and doers, are examples of *strong AI* or *artificial general intelligence (AGI)*. The later, *weak AI* (also called *narrow* or *applied AI*), are focused on single narrow tasks, like winning a game show or a chess match or diagnosing diseases from symptoms.

Another important differentiation is between *symbolic* and *non-symbolic* AI. This lies at the heart of the history of AI. Symbolic AI is all about facts. These facts are written in code and placed within the memory of the AI. When needed, they are recalled. It works on an if-then basis, where if a certain question is asked, then a stored answer is given. Symbolic AI is also called *classical AI* or *good old-fashioned AI* (GOFAI). It is a top-down approach, wherein the software controls what happens, like a director guiding his actors to produce a movie of the director's imaginings. Symbolic AI relies on a giant library of books, with all the answers to whatever the AI unit is tackling. For example, a symbolic AI programmed for fishing would have books on what weight of line should be used, what type of fly is best, a taxonomy of fish, where to fish, how to cast, how to net the fish and so on.

My mother is a brilliant oil painter (not too shabby with acrylics, pastels and watercolors either). She has sold her paintings around the world. I have always wanted to paint like her. I have attended countless art classes in search of the muse. I've read all the self-help, learn-to-paint-in-three-weeks books. I've followed the television programmes. I've hung out with artists, and even teach a course (on ecology and design) in an art college each year. But the muse just turned her back on me and walked away into the ethereal mist. It's not that the muse doesn't like me - muses are above all that nonsense. Rather, I am invisible to her. She is a spirit, and I am mere non-artistic flesh. I just can't paint. I don't have 'it', whatever it is. It's an inherent, non-symbolic thing, not an if/then thing. You can't reduce love, art or life to if/then responses.

Symbolic AI has strengths. Its decisions can be easily understood by humans, as it follows processes laid down in comprehensible steps. It is rule-based, and we make the rules. With the exception of the earliest years and the most recent period, symbolic approaches have dominated AI research. However, it has well-recognized issues. Firstly, in a dynamic, changing, emergent world, the paths of if-then logic can quickly become inadequate. You would need to programme vast amounts of pathways to cover all eventualities. In other words, you have to map out the journey for the AI unit with step-by-step directions, and this can become difficult in new and unfamiliar landscapes. Trade-offs, lack of complete knowledge, rapid changes and stochastic events all mean that the AI may be left high and dry, with inadequate flexibility, crying out *"But you didn't tell me it would be like this!"*

Another big problem with symbolic AI is what has become known as the *common sense knowledge problem*. This problem relates to the fact that our intelligence is not only based on available, explicit facts, but on a whole host of other influences, including cultural, social, historical and philosophical. When we start trying to convert these into if-then statements, we run into an ocean of problems. Common sense is not easily programmed.

Indeed, the relationship between the conscious mind and the subconscious mind is still a subject for conjecture. And in terms of what the subconscious mind actually is, there is very little firm evidence. Symbolic subconscious software doesn't exist. You can't map what you have never seen. Symbolic AI also does not do well dealing with variations in light, pressure, texture and sound.

The symbolic approach became viewed as being too brittle and rigid to fully deliver the kind of intelligence that programmers dream of. And it was not only the dynamic nature of the world that proved problematic, requiring extremely complex calculations to model it, but also the ambiguity. The world is a functioning system, and therefore is an emergent entity. It is non-linear and full of surprises, unless you possess omniscience. While we may feel like gods, making AI in our own image, we lack a number of deific qualities, one of which is omniscience.

Of course, symbolic AI can be very good at some things. Deep Blue, who beat Gary Kasparov in that historic chess contest in May 1997, used symbolic AI. It could consider and assess 200 million moves per second. Deep Blue started life at Carnegie Mellon University in Pittsburgh in 1985, where that smart soda machine lived. Indeed, the graduate students who first worked on Deep Blue in its earliest stages of development, Feng-Hsiung Hsu, Thomas Anantharaman and Murray Campbell, may well have used that very soda machine. Maybe there was something in the soda, but that's a whole other discussion. Later developed at IBM, what would become Deep Blue represented the zenith of AI at the time.

Non-symbolic AI (also referred to as *data-based AI, connectionist AI* or *empirical AI*) operates in a completely different way (note that you will also encounter *sub-symbolic AI*, which many authors equate with non-symbolic AI, though some define as an intermediate technology between symbolic and non-symbolic AI. For example, Willshaw (1994) describes sub-symbolic AI as being built up of entities called sub-symbols, which are the activities of individual processing units in a neural network. We will use the term non-symbolic AI.

The aim of non-symbolic AI is to mimic the human nervous system, and in particular the neural activity of the human brain. This is a bottom-up approach, designing the hardware and then letting it produce the outcomes. The AI unit is fed a stream of raw data, and it then identifies patterns within that data. From this analysis it builds a representation of its world. The best fishermen don't rely on a library of facts alone. In fact, many have never read a single book on fishing. Rather, they have gained insight from countless fishing trips to all sorts of different aquatic landscapes. Big ponds, lakes, oceans, small rivers, swamps – they've done it all. Hot days, wet days, stormy nights and ice have all been faced. Fundamentally they have been exposed to lots of different learning experiences and have learnt how to fish. This doesn't rely on symbols, libraries or if/then decisions. It's non-symbolic.

Non-symbolic AI learns, rather than regurgitates, and excels at pattern recognition and classification. As amazing as this sounds, the approach requires huge amounts of data in order to approach an accurate representation of a real-world problem. Non-symbolic AI is more to do with hardware than software. It uses a series of artificial neurons, connected together, to mimic the structure of the brain. This is called the *connectionist approach* because of the importance of the connections.

Structurally, it consists of large interconnected networks of simple processors that run in parallel. At any one point, a given processor may be delivering input or output, and what emerges is knowledge in the form of a pattern composed of the strengths of connections between these processors. These patterns are products of the activation states of each processor and of the linkage strengths between each processor. This neural network is like a black box. You feed lots of data into it, and something comes out, but you can't follow the thinking and can't be sure how it ended up with the outputs that it did.

An example of a neural network is the *AlphaGo* AI from Alphabet's DeepMind. AlphaGo was the first computer to defeat a professional human Go player, in October, 2015. Go is an ancient Chinese strategy game and is thought to be much more difficult to play than chess, because of the larger board (19 × 19 squares as opposed to 8 × 8 squares in chess) and the greater variety of possible moves. The game is played by two players, one with black stones, who starts, and one with white stones. Players can place their stones on any free point (where two lines cross), but if one of your stones is surrounded by the opposite color on all sides, the stone is removed. The winner is the one who surrounds the most territory and captures the most stones.

At the heart of non-symbolic AI lies *machine learning*. Machine learning represents a watershed in AI development. The AIs work out their own mappings, rather than depending on programmers, using the huge data sets flowing in to them. This has led to major increases in cognitive and perceptive powers. AlphaGo used machine learning, improving each time it played. It was given training data and played against humans and other computers. This ability to learn and to improve is viewed as an important technological advance. It also sets non-symbolic AI at the forefront of technologies.

However, whether it is Go, checkers or chess, these games are examples of what are called *perfect-information games*, where both the human and AI unit know the exact state of the game at any given point. More challenging are *imperfect-information games*, such as poker, where a player can hide cards and bluff. Such games are very much like the real world, where we ultimately want our AI units to function. In the real world there is a lot of hidden information and complexity, partly because of the emergent qualities of complex systems, and partly because of deliberate subterfuge. Such hidden-information games pose useful testing grounds for AI. Recently for the first time, an AI called *Libratus* has beaten top professionals in two-player Texas hold'em poker, the most popular form of the game in the world (Brown and Sandholm, 2018).

The dream of machine learning has been anticipated for many years. Remember the proclamation of Marvin Minsky in 1970 that "*In from three to eight years we will have a machine with the general intelligence of the average human being*" (in Darrach, 1970) or the even earlier proclamation by Herbert Simon (from Carnegie- Mellon University again) that "*machines will be capable, within twenty years, of doing any work a man can do*" (Simon, 1965)".

And the hyperbole surrounding the field in the 1960s has returned. Grace et al. (2017) claim that: "*AI will outperform humans in many activities in the next ten years, such as translating languages (by 2024), writing high-school essays (by 2026), driving a truck (by 2027), working in retail (by 2031), writing a bestselling book (by 2049), and working as a surgeon (by 2053). Researchers believe there is a 50% chance of AI outperforming humans in all tasks in 45 years and of automating all human jobs in 120 years.*" Futurist Raymond Kurzweil predicts a moment of technological singularity, where AI will overcome human capabilities, ushering in a utopia, as predicted by some philosophers in the Enlightenment (Kurzweil, 2005). Robert Geraci (2008) looks forward to a "*victory of intelligent computation over the forces of ignorance and inefficiency.*"

However, weaknesses still exist, particularly in less clear-cut areas of application. For example, the training sets of data can have biases within them. To rule out such bias, a huge amount of training data would need to be gathered. Yet if such biases have gone unrecognized and are fundamental to a particular area of study, this could be difficult to rule against. Also, a game of Go is not real life. There is a fixed set of rules, a finite, albeit large, playing area and a rich data set of previous games to learn from. Novel, dynamic and emergent situations may not be so easy to deal with, even for AlphaGo.

Perhaps the most serious issue with non-symbolic AI is the reliance upon mathematical optimization. The solution chosen is the most optimal one. This is problematic in terms of trade-offs and systems theory, as we shall explore later. At this point, let's just say that sub-optimality is generally the rule in systems, and if you try to optimize any given component, this will ultimately lead to system collapse. Imagine a squirrel, who optimizes ways to find all of the nuts it buried. The outcome would mean there were no nut stashes left, and so there would be no new trees growing from these forgotten stashes. Over the years, this would lead to the disappearance of the very forests that produced the nuts in the first place, and the squirrels would starve. Hence, optimizing for the needs of a hungry, greedy squirrel would not be sustainable.

An interesting way to explore the difference between symbolic and non-symbolic approaches is in the way children learn language. For the first few years, this is an informal process, where children learn through playing a linguistic game with their parents, without any real rules and a constant data stream. Connection strengths develop between neurons. It is a black box approach, where language emerges from the grunts and laughs of a baby, to spoken words, then couplets and finally sentences. This is akin to non-symbolic AI. Once the child starts school, this changes, with rules of grammar now directing things. Spelling, writing and syntax follow. This is akin to symbolic AI.

What is interesting is that both approaches combine to deliver our linguistic ability. In AI, there is now also a move to combine symbolic and non-symbolic approaches. Here, symbolism operates within an overall non-symbolic network (Gärdenfors, 2000; Chella, 2004; Forth et al., 2010; Öztürk and Tidemann, 2014). In what has been called *perceptual anchoring*, the symbols associated with a particular object can be anchored to the sensory data coming from the outside world, in space and time, allowing correspondence between what is sensed in the outside world and what is symbolized. Here, the concepts related to an object (the qualitative, symbol level) can be combined with quantitative data flowing from sensory perception of the outside world.

The respective strengths of each approach are envisaged to combine together, and diminish the respective weaknesses. Rhett D'Souza puts it this way: "*the non-symbolic representation-based system can act as the eyes (with the visual cortex) and the symbolic system can act as the logical, problem-solving part of the human brain*" (D'Souza, 2018).

Historically, both concepts had been developed in the 1950's. William Grey Walter's turtle robots, built in 1951, were early non-symbolic machines. However, the symbolic approach came to dominate in the 1970's and 1980's, receiving most of the funding, and taking the lion's share of research journal pages, graduate programmes and conferences.

The whole field of AI took a huge hit from a report compiled by Professor Sir James Lighthill, entitled *Artificial Intelligence: A General Survey*, published in 1973. This demonstrated that it would be impossible for AI to solve anything but the most basic problems because of the combinatorial explosion or rapid increase in complexity, when operating in the real world. A debate in the Royal Institution in London following the report brought together some of the great minds to discuss the issues raised. It makes fascinating viewing and is available on You Tube (Lighthill Debate, 1973).

There are so many variables and dynamics, that the possible outcomes are close to infinite. As Carl Sagan said, referring to bacterial exponential growth, "*Exponentials can't go on forever, because they will gobble up everything*" (Sagan, 1997). The same applies to computer memory. Lighthill's attack dramatically impacted research funding globally, with governments withdrawing support on both sides of the Atlantic, leading to what was described as the *AI winter* that would last from 1974 to 1980. However, even during the AI winter, the non-symbolic, connectionist approach was viewed as less worthy than the symbolic approach. This was because the non-symbolic approach was thought to need far more calculation and computer power than the symbolic approach.

A re-framing of AI took place towards the end of the 1970's. Instead of promising strong AI, that could ultimately think and plan beyond the capabilities of the human mind, AI was instead marketed to businesses as a tool to help with specific problems, and these tools started to be created and sold. AI became a practical engineering project rather than a theoretical one. The aim was no longer to change the world but to sell some useful software to companies and make a living.

As the vision for AI became tamed, so too were many of the myths slain. AI was no longer a threat that would take over the world and destroy humankind, like some alien, or the dogs of war unleashed. Instead it was that plastic and metal box of tricks, sitting in the workplace, helping with mundane office life. How had the mighty dreams faded? Deep Blue restored some pride, but while being a chess champion was good, it was far from artificial general intelligence.

But, as we know, hope is the last to die, and adherents to the non-symbolic church of the latter-day AI regrouped. Towards the end of the Eighties, a shift began, noted by Paul Smolensky (1988), who wrote "*In the past half-decade the connectionist [non-symbolic] approach to cognitive modelling has grown from an obscure cult claiming a few true believers to a movement so vigorous that recent meetings of the Cognitive Science Society have begun to look like connectionist pep rallies.*" And with its resurgence, non-symbolic AI re-awakened the mythologies of fear and fantasy.

One of the leading advocates was Rodney Brooks, the Australian roboticist, who founded *iRobots*. In a seminal paper written in 1990, entitled *Elephants don't play chess* (this line of attack continued as late as 2010, with the Kelley and Long (2010) paper, *Deep Blue cannot play checkers*), he opened with the unambiguous criticism that "*Artificial Intelligence research has foundered in a sea of incrementalism*", going on to argue that "*the symbol system hypothesis upon which classical AI is base[d] is fundamentally flawed, and as such imposes severe limitations on the fitness of its progeny*" (Brooks, 1990).

Brooks set out a brave new world, where rather than building a representation of the real world as an extremely complex model of overwhelming immensity, you take the approach that this world is its own best model. It is totally up-to-date, and so all you have to do is sense it often enough and in an appropriate way. By tapping into the here-and-now, with a constant data flow, rather than trying to reconstruct it with symbols, you overcome the risks of your model quickly becoming out-of-date and cumbersome.

When we combine the non-symbolic approach with the IoT, we end up with *cloud robotics*, where each individual AI unit is receiving data from their surroundings, learning, and then sharing the learning with all of the other units. This flow of raw data and information has the potential to be extremely powerful.

The symbolic, software approach relied on the assumption that everything was programmable, but it became clear that the real world was much more complicated and ambiguous than had been initially thought. To know anything, the machine first had to know everything. Limited scripts, such as how to win a chess game, did not approach true intelligence. Not only are we unsure how the brain works, but it has become clear that every brain works slightly differently, based on its journey through space and time.

In other words, we possess *mindful brains,* as Edelman and Mountcastle (1978) discussed in their book of the same title. And these mindful brains are emergent outcomes of our social and environmental realities. Furthermore, it is the "unknowingness" that defines the mind, whereas knowingness defines the brain, and cognition becomes a process of filling in the gaps based on our unique experiences. As Wolfe explains (1991), *"Winning games is something our brains do; playing them is something our minds do."*

And so, when it comes to understanding what AI is, we need to embrace both the breadth of intelligence that exists in Nature (due to a lack of species barriers), the individuality of intelligence that exists between us and the connectivity of these intelligences, represented in society.

This brings us back to a fundamental issue in AI philosophy, that sees the mind as a series of components made from other components. These, in turn, are made from other components, the *biochemical mind.* The smallest components are fundamentally not capable of any deep thinking. They may only answer yes or no, like a binary switch, on or off, 0 or 1. This reductionist approach then forms a basis for imagining that artificial intelligence can be likewise assembled.

However, systems theory would disagree, and would claim that emergence, not reductionism, rules, where the characteristics are properties of the whole (including our environmental and societal interactions), not of the parts. In other words, intelligence is an emergent property, not a reductionist one. Here, individuals are outcomes of society and environment, emergent outcomes themselves, with all levels feeding back and interacting with each other and contextualizing our existence and learning. It is within this myriad of connectivity that the survival of the human race itself lies. So how can AI approach this? Well this, ultimately, is what this book is all about.

In the next two sections, we broaden our perspective, examining the contexts and issues that surround AI at present, and, potentially, in the future. We discover that these concerns tell us more about the human race than about the technology itself.

SECTION II

Should I Stay or Should I Go? Ethics in AI

"Jace stood watching Valentine without expression. 'He's a vampire, that's true,' he said. 'But his name is Simon.'"

–Cassandra Clare, *City of Ashes* (Clare, 2008)

Having examined the development and current status of AI, we next address the thorny issue of ethics. If AI is able to participate in decision-making and problem-solving, then what ethical and philosophical frameworks should be used? All decisions we take are informed by the values we hold. Actions have implications, and thus moral philosophy is relevant here. What would such a framework look like, and who should decide this? What if more advanced, sentient machines with artificial general intelligence were able to develop their own frameworks? Depending on which philosophical basis the AGI selected, what could be the implications for us in terms of the decisions made and the actions taken? These are extremely important questions and require urgent attention, given the rapidly expanding role of AI in so many aspects of our lives.

II.1. Choosing an ethical framework

The ethical questions surrounding AI are manifold. If we were to truly create an intelligent machine in our own image, with sentience and emotions, then surely it would have rights? Michael LeChat (1986) asks *"Is the AI experiment then immoral in its inception, assuming, that is, that the end (telos) of the experiment is the production of a person?"* Surely this would then mean that we had an ethical responsibility to treat such technology with respect. Roman Yampolskiy (2013) has gone so far as to suggest that any work on artificial general intelligence, or strong AI, is unethical because *"true AGIs will be capable of universal problem solving and recursive self-improvement. Consequently, they have the potential of outcompeting humans in any domain, essentially making humankind unnecessary and so subject to extinction."*

As an intelligent being, an artificial general intelligence (AGI) may well develop its own philosophical framework within which its actions would have an *ethical framework*. What could this look like? Would it resemble our own ethical framework? And how has such an ethical framework emerged in humans anyway? What is the basis of our morality? Is it an invisible hand, such as described by Adam Smith, in resonance with a broader societal morality (so called *social choice ethics* or *wisdom-of-the-crowd*)? If so, this must rely on a healthy functioning society. In that case, what is a healthy functioning society anyway, and how do we nurture such a thing? Could AI develop within a digital society, and what would such a virtual society look like?

These issues are important, and so we need to have a serious look at the whole subject of ethics at this point in the book. And to do this we need to gain some understanding of what ethics mean for us. Many of us struggle to understand what human ethics and moral judgement actually represent in terms of our own lives. Yet ethics fundamentally inform decision-making and problem-solving, and act as a foundation in terms of justifying our actions and assessing the actions of others. There are a wide range of theories relating to ethics, many of which contradict one another, and this can add to the confusion. We really need to examine at least some of the major schools of ethical philosophy at the outset of this discussion.

Virtue ethics focuses on the individual and their intentions. The flaws and strengths of a person's moral framework will determine the outcomes. In virtue ethics, individuals require moral education to help them determine what is virtuous and what is not. An action is viewed as being morally right if the actor is of morally good character. This has often been couched within the context of actions that lead to the flourishing of the individual or the population. However, defining what *'flourishing'* means is highly contentious, and can largely depend on what school of political philosophy you attach yourself to. A socialist would have a very different opinion than a neo-liberal capitalist.

Deontology has its focus upon whether an action is fundamentally right or wrong. Deontology comes from the Greek word δέον (deon), meaning *duty*. You have a duty to carry out the rules. It doesn't matter what the consequences are. It purely depends on carrying out a set of actions that are written in stone, like the commandments. You should never kill and you should never steal, even if killing someone stopped them killing others. No ifs or buts, it's the rulebook that matters. Deontology resembles the if/then basis of symbolic AI.

Utilitarianism takes the opposite approach to deontology. It is all about outcomes, rather than actions, and we should do whatever it takes to get the best outcomes. So, if we kill the gunman, but save two hostages in the process, this is better than the gunman killing his two hostages. Utilitarianism would say *"take the shot"*. It focuses on the consequences of our actions.

In the *ethic of care*, the answers to moral questions emerge from interpersonal interactions, rather than from the individual. This is very much related to Adam Smith's invisible hand, wherein the individual resonates with the good society and then sympathises and articulates with the good that is observed. Thus, virtuous self-interest emerges. Here, system theory has a significant role to play. This resembles non-symbolic AI, where connectivity is important.

So, we can see that these schools lie on unique paths: actions, outcomes, interactions and scales all differ. Each school requires some decision as to what is good or bad. Who makes this call? And what if there is a difference of opinion? Who adjudicates? When we consider an ethical framework for AI, what do we do? Should we settle on a single school of ethical thought, or explore a diversity of thinking? Do certain situations require particular approaches? It is difficult to envisage how this could work, as each ethical framework tends to exclude other frameworks, based on their primary goals. Each has a well-argued philosophical basis, and has red lines that cannot be crossed.

II.2. The strange case of Asimov's laws

CERNA (Centre for Industrial Economics) has identified three key areas in which ethical considerations are essential in terms of robotics: human-robot augmentation; decision making and autonomy; and social and affective interactions between humans and robots (CERNA Opinion, 2014). Each of these carries particular ethical challenges (Grinbaum et al., 2017).

At present there are just four moral precepts for AI, that, ironically, were born in science fiction, in books by Isaac Asimov, but have now been adapted by governments such as the EU, virtually unchanged (European Parliament Committee on Legal Affairs, 2016). This would appear to be an extremely dubious approach. Many have questioned the suitability of these laws, three of them dating from 1942, with the other, the zeroth law, appearing in 1950, at a time when computing technology was at its very most basic level.

Bostrom and Yudkowsky (2014) have noted that "*If Asimov had depicted the Three Laws as working well, he would have had no stories.*" In other words, these laws were deliberately flawed in order to have conflict and disaster, making a good read. However, it is surely not a good idea to transfer such ideas, in their entirety, into regulatory devices that promise to guide our work on AI, as the EU have done. That seems like a strategy of doubtful value, if not completely irresponsible. Blurring the line between fiction and fact, particularly with something so important, is a stunning piece of stupidity by the EU.

The zeroth law, proclaiming that a robot may not harm humanity, or, by inaction, allow humanity to come to harm, creates difficulties. Firstly, how is humanity defined? Secondly, if some of humanity are destroying the rest of humanity, how much of humanity would need to be destroyed before the robot would step in to prevent harm? For example, in the case of climate destabilization, created by some of humanity but harming all humanity, should AI act against the proportion of humanity harming the rest of humanity? In the case of a nuclear conflict, where many millions of people could be killed by both sides, should robots harm any particular side in order to protect the rest of humanity? In other words, could a robot conceive of a just war?

The first law, that a robot may not injure a human being or through inaction allow a human being to come to harm creates tensions within itself. What if to prevent harm, another human needs to be harmed? This relates to the important but extremely tricky area of trade-offs in ethics. In the so-called *'greater good'* scenario, should an automated vehicle (AV) choose to kill its passenger (and owner) if, in so doing, it saves twenty pedestrians? What if there was only one pedestrian at risk?

Interestingly, when questioned, more people preferred other people's AVs to behave in this way than for their own AV to sacrifice them for the greater good (Bonnefon et al., 2016). Should there be trade-offs at all, or should deontology rule, where standards and red lines are laid down and cannot be crossed? Absolutism is certainly easier, but in system thinking, we must consider trade-offs as a natural consequence of how our complexed, interacting world operates.

The second law, that a robot must obey orders given it by humans except when such orders conflict with the first law, again creates a moral dilemma and also an act of subjugation (obeying the master).

The third law, that a robot must protect its own existence as long as such protection does not conflict with the first or second laws, gives with one hand and removes with the other, and again creates the potential for difficulties in interpretation, already inherent in laws one and two.

It is not the first time that fiction has been employed to form the basis of a factual approach either. Remember the *Red Queen Hypothesis*? It states that organisms must constantly adapt and evolve in order to survive, let alone thrive. It was based on the Red Queen from Lewis Carroll's *Through the Looking Glass*, where Alice has to run fast in order to stand still. The Red Queen says "*Now, here, you see, it takes all the running you can do, to keep in the same place. If you want to get somewhere else, you must run at least twice as fast as that!*"

In evolutionary biology, the concept was meant to demonstrate that because everything else was evolving, any given organism had to evolve too if you were to survive. This has been termed an *evolutionary dance through time*. However, this greatly oversimplifies things, and is highly reductionist, denying emergent properties within systems. For example, Anthony Barnosky (1999) argues that abiotic factors have a much greater impact, setting out his theory, cheekily called the *Court Jester Hypothesis*.

Others, such as Stephen Jay Gould (1990), point out that luck plays the biggest role, in what he terms *contingency theory*. Here, large scale events completely restructure life, and adaptation plays a very minor role. He concluded that if we ran the film of life again, we would have a completely different movie.

The problem with borrowing from fictional literature is that you can over-extrapolate, and, also, that people take it, ironically, as beyond question. *"It must be true because I read it in a fairy tale"*. This can prevent proper examination of the detail of the claim. This is the case with Asimov's four laws of robot ethics.

If artificial general intelligence is truly possible, then surely the AI will have the ability to generate its own moral code, rather than being programmed. If this is the case, what will determine that code and what will be the consequences?

II.3. Free will and moral judgement

Weak AI machines, controlled by algorithms, do not have free will *per se*, and so, at least according to Kant, they cannot possess morality. Moral judgement requires the ability to freely choose, as opposed to the programmed determinism of a robot. Any moral issues, in terms of weak AI, relate to the impact of the algorithm upon other individuals, other robots, other organisms or the environment. Will the output of the programme cause injury, or increase risk?

The more interesting question at this point is this: when is free will truly free?Arthur Schopenhauer, the German philosopher, is of relevance here. He stated that moral freedom represents the opposite of necessity. And, if a robot is programmed, as in weak AI, surely it operates under necessity. Thus, it cannot have moral freedom. And so, at this early stage in the development of AI, the ethics reflect those of the programmer, not the robot. Thus, the robot is an extension of the programmer, and an extension of their moral judgement. If and when strong AI becomes reality, the robot itself will be able to have moral freedom. This is a completely different kettle of fish.

Fully ethical agents possess consciousness, intentionality and free will. But do humans ever truly have free will? We are often programmed by formal education, legislation and peer-pressure to behave in restricted ways. Riots can occur when such controls are removed, symbolic of the veneer that responsible behaviour can be. This veneer can be stripped away, revealing the rough chip board beneath, with little effort. Consider the class of kids with a stand-in teacher, behaving badly as they feel unrestrained, or the impact of alcohol on behaviour as inhibitions are removed.

Are these often ugly or violent displays the result of us climbing off the shoulders of Schopenhauer's blind giant and re-finding our ability to walk, free of will, rather than with free-will, expressing our true, unrestrained, uninhibited and unbridled selves, or are they outcomes of chaos and insanity? What if AI had free will? Could such excursions of madness afflict them if they became drunk on data?

Yet people in a vegetative state, new-born babies or humans with significant mental impairment may not be able to exercise free will and ethical judgement because of brain injury, not yet having a developed brain functionality or brain development issues, but, surely, they are still protected by the Universal Declaration of Human Rights? When does an increasingly intelligent robot gain moral freedom? And how will we know when a machine with AI has achieved sentience or consciousness? What tests can we apply? Just as we cannot know how sentient a cat is, or what its thoughts are, other than by observing its outputs, in the same way, we also cannot be sure that the inner thoughts of a non-symbolic AI robot are.

Roderick Firth, a professor of philosophy at Harvard University, set out what he felt would make the ideal moral judgement in a given situation (Firth, 1952). Firth claimed it would require four characteristics: omniscience, non-biasedness, dispassionateness and empathy. If an AI could have all four of these, then it could be the ideal moral judge. How you would arrive at such a machine is difficult to imagine. Empathy, in particular, could be challenging, and demonstrating empathy while remaining dispassionate would be a difficult trade-off.

In the future, the computer engineer will have to master ethical cognition, or at least have a fundamental understanding of it, in order to ensure safety and to repair functionality if AI activity proves less than safe due to some failure in morality. The engineer may have to become psychiatrist and priest to a morally-perplexed robot with artificial general intelligence.

It may be, of course, that as we see how an advanced AI responds, we become more informed about how we ourselves think. The development and maturation of thinking in a machine may highlight issues for us. On the dark side of this, many psychopaths and predators have escaped notice completely for years, holding down respected positions in society. Dr Harold Shipman, the British GP, practiced as a medical doctor for a quarter of a century before being caught, by which time it was thought that he had murdered two hundred and fifty people. Yet there had been no obvious reason to suspect him until he made the mistake of altering one of his victim's wills. And so, it is not easy to detect significant flaws in the intelligence of a human. What hope might a software engineer have of similarly detecting flaws in a machine with AGI?

Another ethical issue relates to employment displacement by AI. The loss of jobs, incomes and status represent moral issues, with damage to dependant families and society, creating harm and hurt. Thus, the ethical use of AI includes its impact on employment. De-skilling of the workforce could also damage the resilience of the human race if AI were to fail (for example due to a geomagnetic solar storm).

Ethical issues also arise around utilitarianism. How does an AI determine all of the consequences of its actions? If biased or incorrect data is generated and used, leading to faulty decisions, where does the blame rest? In this age of big data, with billions of bits of data being gathered every day, who is checking how accurate each input is? Mistakes can have devastating repercussions.

On 22nd November, 1980, four lobstermen, William Garnos, aged 30, David Berry, 20, Robert Thayer, 22, and Gary Brown, 25, were swept overboard from their boats, the *Fairwind* and the *Sea Fever*, in a storm in the Georges Bank fishing ground, one hundred and fifty miles off the coast of Cape Cod, Massachusetts. Both ships had checked the forecast from the National Weather Service. This forecast relied on data from buoys, ships, high atmosphere balloons, and satellites. The forecast claimed that there was no danger of a storm at their destination, which was much further to the east. They carried on, but were overcome by ferocious winds and waves some sixty feet (twenty metres) high. Their boats sank. It turned out that there were malfunctioning wind sensors on one buoy, Station 44003 Georges Bank, and four young men lost their lives as a result (New York Times, 1984). The accuracy of every piece of data in a big data array can matter.

Immanuel Kant, the German Enlightenment philosopher, in his great work *Groundwork of the Metaphysics of Morals*, originally published in 1785, suggested that we should *"act in such a way that you treat humanity, whether in your own person or in the person of any other, never merely as a means to an end, but always at the same time as an end"* (Kant, 2002). By this he meant that the human should not be used for someone else's gain without the subject benefitting in some way. This means that humans should have consent, experience fairness and have ownership and privacy, while not suffering while being used as a means in terms of, for example, discriminatory profiling.

This could provide a new rule of robotics, replacing the flawed and fictional Asimov laws. It would go something like this: AI applications should treat every human being always as an end and never only as a means.

Data ethics is defined as the branch of ethics that relates to moral problems associated with data generation, curation, analysis and use, with algorithms in artificial intelligence and with practice in terms of liability and responsibility. As the internet of things gathers increasing power and capacity, the age of big data brings with it many ethical issues, as does the encroachment of AI into more and more of our lives in less and less visible ways. Added to this, the acceleration in research and development, mostly in the private sector, means that ethical and legal guidelines are lagging far behind. The private sector is less accountable than the public sector, and tends to publish less in accessible media, for fear of handing any advantage to competitors. This is a potentially dangerous situation.

The challenge now is governance and regulation of the infosphere, not digital innovation. Governance is the development of good practice in management and development, whereas regulation operates through legislation and compliance. And in a fast-moving field, anticipation is key, given that it takes time to develop governance and pass legislation, let alone create an ethical framework in this new landscape.

However, because of the hype surrounding AI, we can become like rabbits in the headlights of a technological eighteen-wheeler (possibly a self-driving one). Yet what we are dealing with, ultimately, is not really that new, because the technology works through existent materials and methods. The fundamental differences are in production efficiency, speed of decision-making and problem-solving, but the outcomes, in ethical terms, are still either good or bad for us as individuals, societies and environments.

Given the existential crisis that confronts us at present, we surely must start from the point of what we want our society and environment to look like, and then work back to what kind of technology can deliver this, rather than the other way around. And this is a key point in this book. Taking control, by steering things, is far better than reacting.

II.4. The confused owl of Minerva: Dangers of a moral vacuum

Georg Hegel, the German philosopher, bemoaned the slow sorry chase that philosophy gave to events, claiming that it was always catching up, like the *owl of Minerva,* who only came out as the day was nearly ending. However, this is a very flawed metaphor, because for nocturnal creatures like the owl, the night is their day, and the day is their night, and so the owl is coming out at the equivalent of its dawn, not its dusk (I'm glad I've got that off my chest because it has always irritated me).

The Hegelian owl of Minerva, if it exists, probably has a saying that goes something like "*Philosophy is always catching up too late. It's like those humans, who get up at daybreak, when all the action has finished. Oh, foolish ones. Twit twoo!*" However as flawed as the metaphor might be, we must ensure that we are pro-active with our ethical, governance and regulatory frameworks, and that we have a clear owl-like vision of what we want our technology to do.

If we desire a good society and great environmental conditions in which to thrive and enjoy our lives, while passing on a healthy planet to future generations, then what better way of doing this than to set sail, along with our technology, in the direction of these values, harnessing ourselves and our technology to the fundamental task of delivering this future. We'll work out how we can do this a little later.

Another huge gap in data ethics relates to children. It is estimated that one third of internet users are children, but there has been a significant lack of legislation, regulation, governance and ethical consideration for this group of potentially vulnerable humans, as they adapt physically, mentally and emotionally in a world so fast-moving that their parents are ill-equipped to help them navigate. As a parent of a twelve-year-old child, I know that I can feel completely at sea in terms of his interactions with the infosphere.

And this is brought into sharp focus because, more generally, we have a completely different set of laws protecting children than we have protecting adults. When a child turns eighteen years of age, they enter a very different legislative world. While we all have the *Universal Declaration of Human Rights,* children have the *Convention on the Rights of the Child,* which addresses issues particularly pertaining to children.

The Convention is then applied in law in most countries and children's social workers endeavour to ensure its adherence as do any professionals working with a child. Yet, given how potentially catastrophic can be the consequences of internet bullying, grooming and exposure to inappropriate material, there is no parallel convention on the infosphere, particularly in terms of collection and analysis of data from children. There surely needs to be urgent action on this, and a detailed convention, specifically focused upon children and the infosphere, must be a priority at an international level, rather than a few extra pages in a more general convention.

One example of the vast change in the lives of children over recent years is the persistence of our digital footprints. While adults are also impacted, during our own childhoods, the internet of things was not a thing, or just a soda machine in Wean Hall, and so there is much less of a digital record of our childhoods, adolescence and young adulthood, thankfully. However, the child today is being constantly mined for data in the big data world, where everything from Alexa to CCTV, and from console gaming to social media is feeding information into the infosphere. And so, when our children reach our age, there will be vast amounts of data on their earlier lives.

While the General Data Protection Regulation (GDPR) of the EU does legislate for erasure of data, the child or parent may not even be aware that the data is held in the first place due to a lack of transparency. There also needs to be some sort of supply chain analysis, wherein the flow of data, its subsequent analysis and the secondary and tertiary uses that this analysis is put to need to be accurately tracked. Also, the GDPR only applies in Europe. Once data diffuses beyond Europe, it is harder to keep track of it.

Robindra Prabhu (2015), from the Norwegian Board of Technology, sums this up as follows: *"Data collection, curation and analysis do not necessarily take place at a single point which can be subjected to robust regulatory measures. Moreover, the technical opacity of algorithms underpinning Big Data analysis, as well as the real-time nature of such analyses, does not easily lend itself to meaningful scrutiny by way of traditional transparency and oversight mechanisms."*

More fundamentally, we need to ask how different the infosphere is from what has gone before. This is important in terms of assessing whether we can merely extend the Convention on the Rights of the Child to cover it, or if we need a completely separate and new charter. Certainly, in terms of big data, global connectivity, data footprints and the invisibility and secrecy of algorithms, there are significant and new challenges shaping the world within which a child develops into an adult. It would seem imperative to begin work on an international convention directly addressing the issues of the infosphere immediately.

II.5. Who's in charge of the big bad wolf?

Algorithms can only do what they are told to do at present, but they can do it very quickly and thoroughly. Thus, any human prejudices within an algorithm will only be magnified, and currently, there is little accountability nor ownership of this process. We cannot blame the creators of these algorithms for the failures entirely. Depending on the training data used, prejudice can come from decisions made many years earlier.

For example, if a programme is trained to determine who should be interviewed for a position based on resumes that are submitted in a recruitment process, the programme may well use previous interview outcomes from previous years to inform its decision-making. If there had been gender bias in previous years, perhaps because of a particular selection committee being male-biased, this would not be detected by the programme, and so it would promulgate this bias. In this case, none of the humans directly involved are complicit in this selection bias, but yet gender discrimination would continue to be practiced.

The matter of who should take responsibility for a particular outcome can create tricky ethical questions, particularly within more complex situations such as in supply chains. If a bag of carrots ends up poisoning people, who is to blame? Is it the person who polluted the farmer's field when he used to own it and had a factory on it forty years ago? Is it this person's grandfather, who opened the factory seventy years earlier? Is it the person who invented the chemical process used in the factory, that produced toxins? Is it the farmer's father, who bought the land cheaply from the person who polluted the field, knowing it was polluted? Is it the supermarket, who bought the carrots cheaply, ignoring warnings from neighbouring farms and failing to check the health of the carrots?

I call this the *murder on the Orient Express problem* of collective irresponsibility. In Agatha Christie's novel, all of the passengers had stabbed a man who had kidnapped and murdered a young girl, but none could be proven to have been the one who had killed him.

The internet of things means that everything is connected to everything else, and so no single AI unit is likely to be easily isolated in terms of responsibility. Rather there will be a form of collective intelligence. Tracing and partitioning responsibility in such a web of interconnectedness would be almost impossible.

An interesting matter relates to who ethics in AI are actually for. Are they meant to allow the machine to act ethically in its dealings and in its decision-making, or are they purely to fit into the ethical framework that the consumer or customer already possesses? Should they be designed to fit into the business ethics of the company or government that controls the stream of data flowing from them, which could well be duplicitous? A company will more likely lean towards an ethical framework that maximizes profits and strengthens its competitive edge. A government may well have an eye on an up-and-coming election.

We want our machines to act in ways which resonate with us. However, with strong AI, the machine is sentient, capable of thinking intelligently and of developing its own moral framework, while networking with other machines that, quite possibly, possess their own alternative ethical frameworks. For surely, moral diversity is as much a part of diversity as any other form of diversity? Are we prepared for this?

I can reflect on my close group of friends in our village. Each of us have very different political philosophies, vote for different political parties and have different ethical frameworks, but that is what makes the group interesting. And so, the idea of intelligent machines with an ethical diversity should not be a challenge, but, rather, an opportunity. But ethical frameworks can be set aside in particular circumstances. We have all read of people who win the lottery and change character completely, transforming from ethic-of-care practitioners to hedonists. Could the same thing happen to our machines, if fed a particular sort of data that turned their heads?

What of lethal autonomous weapons systems (LAWS) in military settings? Do we switch off the ethical framework here, in case it interferes with the ability of the machine to carry out orders? How will decision-making be impacted, and how should the Geneva Convention be interpreted in the design of LAWS?

Of course, before we can envisage what a suitable ethical framework would look like in military contexts for AI, it is worth reflecting that very little open conversation has been held with governments in terms of ethics related to their human soldiers. This may be a better starting point, and may also help in terms of military personnel welfare, both during and after conflict.

This point highlights an issue that is becoming increasingly obvious. Our expectations for AI systems far exceed those for humans, both in terms of responsibilities and ethical frameworks. Can the human in the street, let alone the philosopher, truly define what concepts like justice, risk, consensus, happiness, harm and love really mean? Is everyone working continuously to build capital in these areas? What hope is there of developing an algorithm for such concepts, let alone laying the machine code foundations for AI to develop these from first principles, if we don't already pursue such concepts ourselves?

However, technology can also help in ethical situations. It could scan for bias, check its own training data (a form of reflection) and investigate data acquisition and interpretation, alerting all levels of a supply chain about concerns. And this is really the point: it depends which moral framework we adopt. This is why ethics are so important, both within AI, around AI and in our broader society. Ethics, depending on the particular school we adopt, can hugely impact upon outcomes. Depending on if we base our actions and planning on value ethics or deontology, hedonism (where the primary objective is to maximize pleasure) or utilitarianism will have significant repercussions.

II.6. What should a declaration of AI rights look like?

If AIs attain sentience akin to ourselves and if those AIs are given rights, should these rights include the right to vote and the right to participate in societal ethics? Should AIs be part of the invisible hand, that resonant property of societies that is meant to inform our decision-making through shared empathy? Also, with rights come responsibilities. An AI, if acknowledged to have sentience, surely must also be accountable for its actions in much the same way as humans are. However, what should such rights and responsibility look like? Could an AI claim diminished responsibility in certain circumstances?

Let's look at the Declaration of Human Rights from the UN. Of the thirty articles, issues of gender and sexuality are unlikely to be relevant (although they may be). Would issues of race, colour, language and religion be relevant? Could an AI machine be able to experience a fair and public hearing by an impartial tribunal given the interconnected internet of things? Can an AI be subject to arbitrary interference with its privacy, family, home or correspondence? What will economic, social and cultural rights and free choices of employment look like for an AI? Should AI machines have the right to join a trade union? What about freedom of movement or rights of asylum and nationality? Rights relating to marriage and property are also potentially required.

Human rights are an outcome of the history of humankind, resulting from the challenges and abuse suffered by generations under slavery, suppression and persecution. The historical journey of AI has only just begun, and has not really started in terms of sentient AI. So, there is no historical path along which AI rights can be seen to have developed. If we just sit down and create these rights, then, in terms of a process, this does not resemble anything representative of the creation of our own rights, that have emerged through struggle and proof. Furthermore, the impact of AI is likely to dramatically change the context of both AI and humans, most likely requiring fundamental changes to the rights of both.

In terms of responsibility, as AI technology begins to deliver autonomous learning and development, it will become impossible for the original programmer to predict exactly what the machine will become, and therefore responsibility will begin to shift from the human to the machine. As a university lecturer for many years, I have never been held responsible for the actions of former students who I taught. Yet my teaching could, theoretically, contribute to how these former students have acted.

More fundamentally, should parents, society and government be held partially responsible for criminal acts committed by individuals who were failed by those systems? In much the same way, AI machines that have been trained on data sets, potentially biased in their nature, and initially programmed by programmers with inherent biases, and set to work in contexts inherently biased, cannot, surely, be held completely responsible for biases in their outputs.

Of relevance here is the treatment of children and young adults. The laws related to these two groups differ significantly. On the morning of a person's eighteenth birthday in many countries, individuals wake up in a completely different legal framework than that in which they went to sleep the previous evening. Should this be the case, and if so, should there be a similar shift in responsibility for an AI with developing intelligence, wherein, at a certain point of development, responsibility shifts from the programmer (parent) to the AI (offspring)? Furthermore, not every child develops at the same rate, for a myriad of reasons. Surely the same will apply to different AI units? Should the age of full responsibility vary depending on the individual human or AI unit? The answers to many of these questions will differ dramatically within a deontological or utilitarian ethical framework for example.

Human development marries physiological, physical and cognitive developmental processes, all of which change rapidly during adolescence. Yet robotic AI development is unlikely to be tightly intertwined to such a multiplicity of changes. Hormones are not typically important in AI learning and development. Thus, we must be careful when making parallels between humans and AI. Sentience may be shared, but the vehicles within which that sentience may be seated are likely to be very different. Richardson (2015) reflects that the *"desire to produce human-like qualities in AI machines and robots may shake certain assumptions about what it means to be human."*

It has been argued that ethics cannot be programmed in the form of algorithms, because much of the field of ethics remains unresolved. This is partly down to ongoing philosophical work on the area, on our reliance on intuition, and it is also a consequence of the complex, dynamic contexts within which moral decision-making usually take place. Culture, politics, social issues and life histories also impact on our ethical framework. It is not any more straightforward on the side of AI. With lots of different systems being developed globally, with different 'personalities', motivations and algorithms, any sense of a global ethical system emerging that can work in a fair and even way, while not disempowering individual humans or the human race, seems a particularly hallucinatory pipe dream.

Finally, as in so many aspects of AI, whether it is defining what intelligence is, what a good society looks like, or which ethical framework we should use, by trying to understand the consequences in AI-human interactions, we actually learn so much more about these areas in terms of their application to human-human interactions. In other words, such study may well help us significantly in our understanding of our own place in the world and how we act and react within our social contexts. This has, surely, got to be a good thing. Not only should ethics become part of any information technology curriculum, but it also should be part of the more general school curriculum.

In whatever way that we think about machine ethics, it is surely an unavoidable observation that as AI becomes stronger and we wake up more fully to the infosphere that we increasingly inhabit, technology is making more decisions for us and these decisions have an increasing likelihood of impacting upon us. Thus, we need to explore how to imbue technology with some form of ethical awareness in order to inform this decision-making. To do otherwise would be remiss. Rosalind Picard, founder and director of the Affective Computing Research Group at MIT, puts it like this: "*The greater the freedom of a machine, the more it will need moral standards*" (Picard, 1997).

And it is not just a matter of outcomes. Fundamental to our acceptance of AI technology within our daily lives is the issue of trust. Central to this are transparency, accountability and dependability. Only upon these foundations can a good AI society be built. A functioning society requires resonance, and resonance requires trust. And at the heart of this lies sympathy and moral sentiment, as Adam Smith emphasised. Yet this trust has already been undermined, and next we take a look at some of the great failures in technology, and what they mean for us and for AI.

SECTION

III Gender, Race, Culture and Fear

"Those of us who stand outside the circle of this society's definition of acceptable women; those of us who have been forged in the crucibles of difference – those of us who are poor, who are lesbians, who are black, who are older – know that survival is not an academic skill... For the master's tools will not dismantle the master's house. They will never allow us to bring about genuine change."

–Audre Lorde, *Sister Outsider* (Lorde, A. 1984)

Diversity, as we have already emphasised, is central to sustainability, both in Nature and in society. In humans, gender, race and culture not only bring a beautiful and diverse spectrum of thinking and expression to the table, but have, throughout our tainted history, been the targets of segregation, suppression, hatred and sectarianism. In non-symbolic AI, you are what you eat. The big data on which these programmes feed and upon which machine learning builds its models, are often riven with biases. And the consequences can potentially realize themselves in the decision-making outputs. Concerns around what kind of thinker AI is designed towards are also important to address.

There is a perception that AI is based on male, white, western intelligence, acting as the master's tools, and that inherent bias will impact gender, race and culture negatively as a result. Kate Crawford, director of the AI Now Institute at New York University, wrote that *"Like all technologies before it, artificial intelligence will reflect the values of its creators"* (Crawford, 2016). We now examine the evidence for this, and what lessons we need to learn. Finally, we explore the issue of fear of what this technology might bring to us, from unemployment to the destruction of the human race. How justified are such concerns, and why do we have a tendency towards fear and loathing when new technology evolves?

III.1. Gender issues in AI

The very earliest imaginings of AI as a concept appear very gender-biased. In Homer's Iliad (XVIII, 372-379; 468-473), as we have noted already, the golden maidens created by Hephaestus were automatons that could predict what their masters wanted without having to be told, and to carry out these requests obediently. The obsequious roles of these female automatons, quietly going about their duties for the gods, is unequivocally gender-biased, a fantasy for males to enjoy.

Meanwhile, in the Liezi, written by Lie Yukou, a male automaton was seen to flirt with the king's concubines. Other Chinese work describes hydraulic automatons in the forms of singing girls, all on board a wine boat, in more gender-biased imaginings (Ambrosetti, 2011).

But this is not just the ancient history of fictional literature, whose authors were exclusively males. In AI today, there is a growing unease at male bias. This is undoubtedly partly related to the fact that most people working in AI research and development are males. A self-fulfilling prophecy, this seems to stem from the shortage of officially recognized female role models in AI. They exist, but have been hugely under-reported.

Another issue is that because males occupy most of the senior positions in authority, and are over-represented in big data as a result, then the training data that AI systems feed upon are gender-biased also. In technology as much as in our own lives, you are what you eat. The bias is further amplified by the fact that most electronic personal assistants (such as Cortana, Siri and Alexa) and chatbots (such as Mitsuku, Evie, Tay and Anna) are female, just like the golden maidens of Hephaestus.

Emilia Gómez Gutiérrez (Gómez Gutiérrez, 2019) identifies a number of gender biases in AI. Speech and voice recognition systems performed better for men than for women, contributing to an alienation of female users. Facial recognition systems produce more errors with female faces (Klare et al., 2012). Societal stereotypes of gender permeate search engines, with searches on work producing more male images and those on shopping producing more female images. A similar experience can be had on Google Image Search. If you enter the words 'president' or 'primeminister' into Google Image Search, around ninety five percent of the resulting images are of men (Bano, 2018). Imagine a child carrying out such a search. There could be no stronger example of gender stereotype reinforcement. Of course, we can't just invent lots of female presidents and post their pictures on Google. But we can be aware of the potential impact of such searches and the need to discuss the issues around them.

Another example saw an employer advertising for an employment opportunity in a male-dominated sector via a social media platform. The advertising algorithm on the platform only posted the position to males, in order to maximize returns on the number of applicants (Scheiber, 2018). This is because a search of the previous interview outcomes revealed that mostly males were appointed, and so to 'optimize' the recruitment process, the technology decided to only target males. The decision was not gender-biased in itself, but, rather, was driven by an efficiency emphasis within the algorithm. However, the outcome excluded females from applying.

Recruitment tools that utilize *text mining* or *text analytics* (the process of examining large collections of written resources to generate new information) also demonstrate gender bias. Another situation arises where women who marry often change their surnames to that of their partners. In this case, their digital record can become disconnected from their current identification, causing countless problems within utility services based on big data.

A White House report focusing on big data concluded that gender biases were mostly unintentional but stemmed from perpetuation and promotion of historically embedded biases, poorly selected training data, outdated or incomplete data and decision-making systems based on the concept that correlation always implies causation (Podesta et al., 2014). However, the issue becomes extremely serious when we reflect that what AIs are doing is learning human biases and then turning them into seemingly objective truths.

Computers are more trusted than humans, because we are trained to think of them as logical and neutral. How can binary code act immorally? This is why internet scams have been so successful. This trust extends to biases inherent in the systems because of test data stemming from traditional axes of oppression such as male-female power imbalances. Whereas we will call out such oppression when voiced by another human, we are often unaware of it in our interactions with computers.

It isn't only the gender bias within training data that is a problem. Visual appearances are also re-enforcing gender inequality. Military robots are designed to look like men, whereas companion robots mostly look like women.

Another issue relates to the fact that just twenty nine percent of AI research positions were applied for by females (Shoham et al., 2018), while only five percent of senior management roles in the sector are filled by females (Baker, 2018). *Preparing for the Future of Artificial Intelligence* is a recent report from the National Science and Technology Council that states that the shortage of women and minorities represents *"one of the most critical and high-priority challenges for computer science and AI"* (NSTC, 2016).

And the evidence shows that gender-specific recruitment differences begin at school. Baker (2018) also reports that while eighty three percent of males study STEM subjects at schools, only sixty four percent of females do so. At universities, STEM subjects are studied by fifty two percent of males, while only thirty percent of female students take such degrees.

A mere eighteen percent of computer science graduates are female. What is more alarming is that the situation is deteriorating. In 1984, thirty seven percent of computer science graduates were female (Corbett and Hill, 2015). Less than twenty percent of faculty positions are filled by females in leading universities (Shoham et al., 2018). Only twenty-eight of the two hundred and three countries globally have a female in charge of the government ICT ministry, or its nearest equivalent (Sey and Hafkin, 2019).

At a more fundamental level, many countries have gender inequality in accessing web-based computing. Surprisingly, some of the greatest levels of access inequality are in Japan (nine percent fewer females than males with access to computers) and South Korea (eleven percent fewer females than males with access to computers), given that these two nations are leaders in technology (Sey and Hafkin, 2019). Three hundred and twenty-seven million fewer women than men use mobile internet, while one hundred and eighty-five million fewer women than men own a mobile phone. In South Asia, women are twenty six percent less likely to own a mobile phone than men.

A number of issues are known to exist in terms of overcoming gender inequality globally. These differ across the planet. Infrastructure availability and cost can be problematic. Interest and perceptions of relevance can also be an issue. In Africa, often, local tribal languages are not available on the internet. Cultural issues may play a role, particularly in countries where women are not given equal rights. Finally, even if female users can access the internet, what awaits them can be very negative. Seventy three percent of women online have experienced some form of cyber violence (Sey and Hafkin, 2019).

In the recent *Global Gender Gap Report* by the World Economic Forum (WEF, 2018), the overall gender gap (based on the four pillars: economic opportunity, political empowerment, educational attainment, and health) will be closed in sixty-one years in Western Europe, compared to one hundred and fifty three years for Eastern Europe, Central Asia, the Middle East and North Africa. While this lag prevails, and continues to impact on machine learning within AI systems, then it will be difficult to completely remove gender bias from within these technologies.

Imagine trying to write an algorithm that asks the AI to ignore very specific bias, but at the same time to learn about what society and individuals are thinking in order to build bridges between the human and the machine, especially when these bridges play an essential part in so much of the more advanced applications that AI may be useful for. Filtering will be necessary, but much of the sexism is hidden beneath layers of apparently innocent information, making it difficult to identify and discard. Even if sensitive fields of data on, for example, race and gender are hidden from learning algorithms, it has been reported that these algorithms can reconstruct such fields and then use these probabilistically inferred constructs in discriminatory ways (DeDeo, 2015; Feldman et al., 2015).

Basic statistics, such as the number of females working in a particular sector, may not be sexist *per se*, but, as we have seen, AI could interpret that as evidence to continue targeting males with job offers of a role in that sector as part of its aim to attain its goals efficiently. Amazon have stopped using an AI recruiting tool that favoured male applications (Dustin, 2018).

Melinda Gates (Co-founder of the Gates Foundation) and Fei-Fei Li (chief scientist for AI and machine learning at Google) have founded *AI4All*, which works with ninth grade students from under-represented groups (for example school girls and African American students) in order to address the inequalities in representation. However, the challenges are still significant. Even if we recruit more women into AI, there is still the issue of inequalities in the big data on which AI feeds.

The point is that artificial intelligence is firmly rooted in the real world, with all of its imperfections. Its foundations are not laid in utopian clay, whose very essence is some Garden of Eden, full of only good things. Just like Dean Jocelin in William Golding's *The Spire*, struggling to accept that the sacred cathedral was being constructed by drunken, licentious and, ultimately, murderous workmen, so we must understand that AI feeds of the imperfections of a world plagued with inequality. This doesn't mean that this situation is in any way acceptable. But recognizing it is the first stage in addressing it.

When assessing the impact of new technology on employment, there are conflicting assessments as to what kind of impact AI could have at a gender level. For example, in Bolivia, the automation risk varies between two to forty one percent of jobs being lost, while in Japan the impact on jobs has been assessed as anything from six to fifty five percent (World Bank, 2019). However, most research points not only to humans being replaced by machines, but to there being a gender imbalance in this process.

A recent IMF report (Dabla-Norris and Kochhar, 2018) has shown that women are more likely to face unemployment due to automation than men, and the gender disparity is greatest if aged over forty or working in clerical, sales and service sectors. Particularly vulnerable in these sectors are women between sixteen and nineteen years of age. Furthermore, around fifty percent of women with a high school qualification or less are susceptible to losing their jobs compared to forty percent of men.

Hamaguchi and Kondo (2018) showed that gender disparity increases within increasingly large cities in Japan, with men being less effected and women being more effected. The reason for this is that there are more management-level positions in larger cities, which are more likely to be filled by men. These decision-making positions are much less likely to be automated than the clerical positions, traditionally filled by women, which also increase in larger cities.

While there is the promise of job creation in the ICT sector, women are not well placed to take advantage of this. The data from OECD countries show that only one and four tenths of a percent of female workers have jobs developing, maintaining, or operating ICT systems. This compares to five and a half percent of male workers (OECD, 2017). In terms of gender issues in the workforce, in the USA, only twenty five percent of computer science professionals and fifteen percent of engineering professionals are female (US Department of Labor, 2016). A similar gender gap exists in other countries. In the UK, only sixteen percent of the employees within the IT profession are female (WISE, 2018). Thus, male workers are much better positioned to benefit from the upside, while females are very much worse positioned to suffer from the downside of the technological boom.

It is a further irony that the new creative spaces, called hackerspaces, where people can come to improve on their digital skills and develop projects, are commonly referred to as *Men's Sheds*, not exactly welcoming for women, and stereotypical. Data, albeit from 2012, shows that only ten percent of users of these facilities were females (Moilanen, 2012).

Are male and female ethics different, or, more fundamentally, is female and male intelligence different? If so, should there be two AI sexes, or just one? And what about LGBT genders? Stanford psychologists Eleanor Maccoby and Carol Jacklin, in one of the first studies of gender differences in cognition, temperament and social behaviour, found that males were more assertive and females were more prone to anxiety (Maccoby and Jacklin, 1974), but otherwise, there was little difference between genders in, for example, analytical cognition and higher-level cognitive processing.

Feingold (1994) found that females were more extravert, trusting and nurturing than men. In terms of cognitive ability, this is related to modernity. In the USA, females from younger generations have a higher cognition than young men, while in China, in younger generations there is no cognitive gender disparity, whereas in older generations in both countries, males have higher cognition scores than females (Lei et al., 2012). Furthermore, recent research shows no differences in mathematical ability between genders (Hyde, 2016).

As we have seen in the previous section, moral decision-making may represent one of the most difficult aspects of AI to develop, and at its heart lies the question of what to base it on. What should a morality algorithm look like? In terms of moral judgement, a study by Rebecca Friesdorf and colleagues (Friesdorf et al., 2015) has shown that females tend to use deontology (morality being framed by whether the action is fundamentally right or wrong, based on a fixed set of rules) while males tend to use utilitarianism (where morality is framed by the consequences of the action).

To understand the difference here, let's take an example. An AI working in a hospital realizes that there is not enough of a particular drug to supply three patients with. There are only ten milligrams of the drug left. All of the patients are the same age and gender. However, one patient needs ten milligrams to survive whereas the others only require five milligrams each. Deontology would insist that you cannot remove the drug from the one patient to save the other two, as you will contravene the moral 'rule' of not acting in such a way as to kill the patient. Utilitarianism would insist that saving two of the three patients is better than saving one, and so the drug should be given to these two patients. Which algorithm, if either, should be placed in the hospital AI?

Carol Gilligan framed gender differences in moral decision-making in terms of a male *ethic of justice* and a female *ethic of care* (Gilligan, 1982). This led to the concept of the *Feminist Ethic of Care*. However, questions remain as to how clearly evidenced such gender differences are (see Jafee and Hyde, 2000). Certainly, research indicates no gender difference in cognitive abilities, but in terms of affective responses (psychological responses), females show stronger results. Affective processing is more likely to lead to deontological decision-making, whereas cognitive thinking leans towards utilitarian outcomes.

Fumagalli et al. (2010) and Friesdorf et al. (2015) reported just this outcome, wherein males used utilitarianism in moral decision-making situations involving harm, as opposed to females, who tended towards deontological outcomes, driven by a stronger affective response to harm. There were no differences in cognitive abilities. In non-harm decision-making scenarios, there was no gender-related difference. Given the gender bias in programmers, there could be a risk of only producing utilitarian algorithms, with significant implications for decision-making. This needs very careful research and scrutiny, given the complexities of problem-solving in a dynamic world.

Margolis and Fisher (2002) point out that "*Women must be part of the design teams who are re-shaping the world, if the re-shaped world is to fit women as well as men.*" They further point to an example where "*Early voice recognition systems were calibrated to typical male voices. As a result, women's voices were literally unheard*".

Is there a bias in AI algorithms because most of the programmers are male? How different would AI be if all of the IT sector were female? This is a difficult subject and an acknowledged minefield. If we wish to claim equality, then we would say that males and females are identical with no gender-based differences in abilities. Yet if we stress the importance of females bringing something different and important to the table, we emphasise a gender difference.

Certainly, we have seen growing evidence for gender-based differences in ethical decision-making, but are there other differences, and how different would AI be if these differences are brought to the table in terms of programming? Is there such a thing as a gender-based difference in intelligence and if so, how should this be represented in AI? This issue came into the spotlight in 2017 at Google.

In 2017 one of Google's then engineers, James Damore, released a now infamous internal memo, entitled the *Ideological Echo Chamber*, that found its way into the public domain. In it, Damore claimed that male/female disparities in workforce are partly explained by biological differences, stating that women are more neurotic. The evidence that Damore cited was extremely flawed. While there can be small differences between male and female populations in higher order personality traits, there are much greater differences within each group. In other words, it is wrong to stereotype genders. Also, there is a complete lack of mention of LGBT individuals in most gender research related to such traits, and, in particular, relating to technology roles.

Donna Haraway, in her essay entitled *A Cyborg Manifesto* (Haraway, 2000), writes of the hopes of a new technological age liberating us from gender dualism: "*Cyborg imagery can suggest a way out of the maze of dualisms in which we have explained our bodies and our tools to ourselves. This is a dream not of a common language, but of a powerful infidel heteroglossia. It is an imagination of a feminist speaking in tongues to strike fear into the circuits of the supersavers of the new right. It means both building and destroying machines, identities, categories, relationships, space stories. Though both are bound in the spiral dance, I would rather be a cyborg than a goddess*".

Yet the hope that AI could somehow forge a new, more equal and united future, where differences were not the issue, but rather that a neutral technology could lead the way to tolerance, peace, empowerment and fairness, seems much less likely as time goes by. Bogost (2015) highlights how our culture ascribes an aura of objectivity and infallibility to algorithms. Yet the real world, upon which our algorithms act as mirrors and amplifiers, represents inequality, bias, tribalism and greed. A classic example of this was IBM's AI, *Watson*, that, infamously, had developed an excessive swearing habit from data it had consumed that had to be corrected (Madrigal, 2013).

III.2. Racial issues in AI

One of the most significant and shocking moments in terms of issues with algorithms hit the headlines in 2016, in a study by Julia Angwin and her team. The paper examined Northpointe's Correctional Offender Management Profiling for Alternative Sanctions (COMPAS), that uses an algorithm to score a defendant's risk of re-offending, risk of violence and risk of not appearing for trial. Their findings were startling. This software is commonly used to assist judges in sentencing and parole boards in parole hearings.

Angwin and her team found that black defendants were twice as likely as non-black defendants to be wrongly classified as being of a higher risk of violent recidivism, and white defendants were wrongly classified as being of low risk sixty-three percent more often than black defendants (Angwin et al., 2016). In other words, this algorithm was showing negative bias towards black individuals and positive bias towards white individuals. Another example of racial bias occurred with packages ordered using Amazon Prime, the same-day-delivery service. ZIP codes on addresses from black neighbourhoods in Atlanta, Chicago, Boston, New York, Dallas and Washington, D.C. were labelled as undeliverable (Ingold and Soper, 2016).

Many more very unacceptable race issues arise in facial and image recognition systems, mostly because test data is predominately of white males. Google's photo app labelled black people as gorillas, Hewlett Packard's web camera software couldn't recognize people with dark skin pigmentation while Nikon's camera software interpreted still images of Asian people as blinking (Crawford, 2016). These findings are concerning as facial recognition software is being used more often in diagnostic situations in healthcare, for example, in detecting melanomas. If the software cannot adequately analyse more pigmented skin, it could, in principle, mis-diagnose patients from certain ethnic backgrounds.

Facial recognition software used by American police departments has been shown to misidentify people of colour (Klare et al., 2012). Thus, you are more likely to be mistakenly arrested if you are black than if you are white. PredPol, a set of software used by police departments to maximize efficient distribution of resources by predicting where crime is likely to occur, has led to a positive feedback effect, wherein more police were sent to majority non-white neighbourhoods, thus increasing the likelihood of crime being detected, and this in turn led to greater resources being poured into these areas (Reynolds, 2018).

In the recent Interim report of the Biometrics and Forensics Ethics Facial Recognition Working Group for the British Government (Hallowell et al., 2019), significant concerns were raised relating to the accuracy of live facial recognition (LFR) technology and its potential for biased outputs and biased decision-making on the part of system operators. The report also raised concerns over the lack of independent oversight and governance of the use of LFR.

Microsoft's AI chatbox, called *Tay*, spent a day exposed to Twitter feed. As a result, it started making antisemitic insults. On 24th March, 2016, Tay tweeted *"Bush did 9/11 and Hitler would have done a better job than the monkey [referring to President Obama] we have now. Donald trump is the only hope we've got"* (Hunt, 2016). Shocking stuff, but unfortunately reflecting the Twitter feed of that day. This also resonates with Godwin's Law, that, in one version, states: *"As an online discussion grows longer, the probability of a comparison involving Nazis or Hitler approaches 1"*.

The problem is that it can be very difficult to prove an algorithm has unfairly judged you because more recent algorithms employ deep learning, with unsupervised learning framing the future of AI systems. Such bottom-up approaches, where the thinking is fundamentally more autonomous, mean that the end point or output cannot easily be mapped to a particular pathway.

Weak AI is model-based and follows a logical pathway. Strong AI is like something emerging from a dense fog – you cannot tell where it came from or how it ended up here, even though you know it must have come from somewhere. However, Cathy O'Neil (2016) notes that it is the most vulnerable in society who will suffer the most from biased AI, the poor who can't afford a lawyer to challenge the algorithm-based prejudice, and whose university funding application is turned down because of their postcode.

There is a significant problem in terms of adequate racial representation in the major computer technology companies. For example, according to the Google Diversity Annual Report 2018 (Brown, 2018), only two and a half percent of Google employees are black (compared to a USA population that is twelve percent black), while fifty three percent are white and thirty six percent are Asian.

In leadership positions, only two percent are black, fifty six percent are white and thirty eight percent are Asian. Coupled with this is the fact that more black employees are leaving than any other race, in spite of the employment pool of black employees being so much smaller than white or Asian pools (Brown and Parker, 2019). Similar numbers are found in other leading tech companies. This is very worrying, reflecting a worsening trend rather than one that is improving.

Gender, racial and cultural discrimination represent forms of tribalism, where one group identifies with itself and strengthens that identity by excluding other 'tribes'. It is a deep and polarizing theme that runs across religions, sports supporters, academic schools of thought, the T-birds and the Scorpions,and politics.

In fact, no sooner are we drawn to an interest or subject than tribal traits begin to manifest themselves. The tribe can be united by common ground, so called *in-group favouritism* (Yamagishi et al., 1999), as typified in the song written by Billy Page and made famous by the Mammas and the Pappas, *I'm in with the in crowd*. On the other hand, the tribe can be bonded by *out-group prejudice* (Fiske, 2000), so powerfully highlighted by Nina Simone in *Mississippi Goddam*.

Indeed, shared prejudice can unite a group more than mutual interests. It's ugly and it's nasty. People get killed. Yet it is part of a deep-set theme, so deep that it is difficult to eradicate. In order to maintain diversity, we need some sort of identity within different groups. Without some degree of tribalism, we cannot maintain diversity.

One of the great challenges is how to maintain diversity while avoiding the worst forms of tribalism. A recent piece of research highlights a shocking reality: that groups of robots can develop tribalism without being told to. The research was done by scientists from Cardiff in Wales and MIT in Boston (Whitaker et al., 2018). Roger Whitaker and colleagues ran thousands of simulations of a game of give-and-take, where one hundred virtual agents (represented by algorithms), made decisions of whether to give to someone inside their group or outside their group, based on the receiver's reputation and their own levels of prejudice. The simulation was run five thousand times. The virtual agents either copied strategies within their own group or from the entire population, in order to maximise short-term gains.

What emerged was fascinating. Small groups of agents formed, giving to each other but not to agents outside of their groups. Those outside of the core groups then formed their own groups, fracturing the population. This prejudice was not an outcome of being fed real-world, prejudiced data full of human biases, but emerged within the autonomy of the game, completely independently. They found that such widespread emergent prejudice is hard to reverse. This is a disturbing discovery, for even if we find a way to filter the data that AI systems feed off, in an attempt to stop human prejudices infecting the technology, there appears to be a natural drive towards this sort of division.

III.3. Cultural issues in AI

Culture is a difficult thing to define. It can be linked to particular nations, races, politics, religions or gender identities. It usually involves some form of tribalism. In something like indigenous culture, it can straddle many different geographical and historical contexts.

Culture is an emergent entity, and historical contexts can have significant impacts on its shaping and form. However, one thing that is universal is that culture is fundamentally about the way we interpret the world around us. Communication is not just about what is said but also about what is heard. As such, cultural differences complicate communication. And this is where the Western, male, white dominance in technology production and design, producing a single type of message, represents a cultural bias.

It has forever been the same. There are two possible outcomes. One is that we unify around a single white, Western male culture and everyone joins this monistic order. This denies the importance of cultural diversity for social sustainability, and therefore represents a strategy of doubtful value. Alternatively, AI should become a *pluriverse*, able to communicate and think in the way each user thinks, criss-crossing cultural boundaries while strengthening cultural diversity.

This is a really important and significant issue. For a sustainable future, AI cannot be neutral. It must positively encourage diversity, be it cultural, racial, gender-based or environmental. Because diversity represents resilience, a key characteristic of sustainability. Yet, as we have seen so far, there is a danger that the opposite will occur, built on bias in programming, bias within big data and emergent bias within autonomous systems.

Yet many of the answers are already known. Reflecting back on the work by Whitaker et al. (2018) involving tribal prejudice, they found that prejudice between groups drove cooperation within groups. This is an interesting outcome, where opposing forces help sustain each other. Prejudice strengthens groups. However, strong groups working in isolation weaken overarching social cohesiveness. It comes back to the challenge of maintaining a diverse but functional society.

Thus, things are not simple. There are complexed interactions between all, as we would expect in a complexed system. However, another very relevant finding of Whitaker's work was that the greater the number of groups, the greater cooperativity and the less prejudice there was. Furthermore, societies with higher out-group interactions and global learning again had higher levels of cooperation across the system and had lower levels of prejudice. These findings are extremely enlightening in terms of how to maintain diversity without the damaging aspects of tribalism.

Indigenous culture places an emphasis on society and environment rather than the individual, and on a pluriverse rather than a universe. Rather than the Western developmental programmes, both sustainable and otherwise, where the emphasis is on converting the world into a globalized version of the most powerful nations, pursuing capitalism and individualism in a neo-liberal context, plurality celebrates diversity and sees it as central to sustainability. This includes different economic, social and environmental approaches, engendering a resilient planet and informed by localism.

Such post-development approaches actually pre-date development and colonialism, and were almost lost in the great tidal wave of Enlightenment transition. The pluriverse is a place of opportunity and expression, rather than competitive exclusion and suppression. It represents an ecological approach rather than a Darwinian approach, wherein diversity is maintained and underpins a functional biosphere, rather than being reduced by intense selection.

Pluralism represents a unity between the rest of the biosphere and ourselves, in that while interconnected, our differences are also our strengths. It all begins with landscapes, both physical and metaphysical, wherein the cultural, geological, and aquatic all inform each other and set the context of each interactive ecosystem. Humans are part of these ecosystems. Ecosystems don't compete, annihilate and replace each other, but rather they exist in their particular contexts.

The concept of globalization and development is a nonsense in ecology. The rainforest doesn't try to re-educate and re-order the natural economics of the desert to make it resemble a rainforest. It just wouldn't work. Yet both ecosystems are crucial at a planetary level. As Conway and Singh (2011) put it: "*Notions of the pluriverse imply multiple ontologies, multiple worlds to be known — not simply multiple perspectives on one world. Universalist discourses and globalist projects are grounded in a unitary ontology and imperialist epistemologies which assume that the world is one, that it is knowable on a global scale within single modes of thought, and is thus manageable and governable in those terms.*"

The pluriverse is the natural way of systems and does not need to be created, but, rather, it requires nurturing. Thus, if we are to embrace this approach, it is a matter of recognizing rather than enforcing it. The dominant Western Enlightenment vision circles around individualism, separation, homogeneity and reductionist thinking. Yet our only hope for a sustainable future requires us to operate within the planetary system, and this necessitates pluralism, integration, heterogeneity and holism.

Querejazu (2016) identifies four essential elements for a thriving humanity: *relationality*, wherein social, environmental and economic arenas are interconnected; *correspondence*, where elements are correlated in a balanced way; *complementarity*, where opposites complete each other and become whole; and *reciprocity*, representing the fundamental idea of justice in every interaction. Three types of justice have been pinpointed by Fischer et al. (2012):

1. *Distributive justice* (has there been equity and equality in the outcome?);
2. *Procedural justice* (is the process effective, fair and inclusive?);
3. *Interactional justice* (how our people are treated during the process?).

Later, we shall see how AI is perfectly positioned to deliver a sustainable future if used appropriately, but is also perfectly capable of destroying such a future. The choice is ours, and is probably the most important decision that our generation will make if the sins of the fathers are not to be paid for by future generations. What is really interesting is that in our journey in this book so far, we have begun to see how AI has shone a light on the many different aspects of the human condition, and on the inherent interconnectedness of the Earth system within which it must operate. Dynamic and emergent characteristics of the real world form the greatest challenges for the infosphere.

This is what a pluriverse looks like. It is tied together by narrative intelligence, myths and dreamworlds, that represent interconnectivity across the complex whole, wherein landscape, culture and Nature interact, not in any way that we can represent in cold empiricism, but rather in sub-symbolic ways, much like strong AI.

III.4. Fear and loathing in AI

Along with climate change (as represented in movies such as *Waterworld* and *Mad Max: Fury Road*) and social division (*The Hunger Games, V for Vendetta* and *Elysium*), the fear surrounding AI running out of control has become a dominant theme in the dystopian movie genre. Super-intelligent robots have populated science fiction for many years, from the *Maschinenmensch* in Fritz Lang's 1929 horror, *Metropolis*, to *Hal*, in Stanley Kubrick's *2001: A Space Odyssey*.

The authors and academics, Raffaela Baccolini and Tom Moylan, have noted that "*the dystopian imagination has served as a prophetic vehicle, the canary in a cage, for writers with an ethical and political concern for warning us of terrible socio-political tendencies that could, if continued, turn our contemporary world into the iron cages portrayed in the realm of utopia's underside*" (Baccolini and Moylan, 2003).

Much like protest songs, movies hammer out a warning, but also can bring terror where there is none. Remember *Jaws*? This movie sensationalised and stereotyped the shark as an evil, malevolent force of significant threat to human kind. Yet sharks claim an average of ten human lives each year, compared to eight deaths every day caused by people using their mobile phones while driving in the USA alone. Crocodiles kill one thousand people annually while dogs kill twenty-five thousand humans each year (Denny, 2017). Yet humans kill seventy million sharks per year for food and sport.

However, concerns related to AI are not only found in the realms of fantasy. Some of the most celebrated minds of our time have painted near-apocryphal images of a future where machines out-think and displace humans. The late, great Professor Stephen Hawking warned that AI "*would take off on its own, and re-design itself at an ever-increasing rate. Humans, who are limited by slow biological evolution, couldn't compete, and would be superseded*" (quoted by Cellan-Jones, 2014), adding further that "*Unless we learn how to prepare for, and avoid, the potential risks, AI could be the worst event in the history of our civilization. It brings dangers, like powerful autonomous weapons, or new ways for the few to oppress the many*" (quoted in Clifford, 2017).

Hawking's concerns were echoed by Elon Musk, the founder of Tesla Motors, who, in 2014, said "*I think we should be very careful about artificial intelligence. If I had to guess at what our biggest existential threat is, it's probably that*", adding "*With artificial intelligence we're summoning the demon. You know those stories where there's the guy with the pentagram and the holy water, and he's like – Yeah, he's sure he can control the demon? Doesn't work out*" (quoted in Graef, 2014).

Bill Gates, co-founder of Microsoft, has commented that "*I am in the camp that is concerned about super intelligence. First the machines will do a lot of jobs for us and not be super intelligent. That should be positive if we manage it well. A few decades after that though the intelligence is strong enough to be a concern*" (quoted in Mack, 2015). Meanwhile, Apple co-founder, Steve Wozniak, claimed "*The future is scary and very bad for people*", adding "*Will we be the gods? Will we be the family pets? Or will we be ants that get stepped on? I don't know*"(quoted in Holley, 2015).

It's a theme that runs further back than you might suspect. The idea of humans taking on the mantle of creator, displacing God, but then being undone by the flawed work of their own hands, runs back far into our historical fabric.

But where do such fears come from, and how realistic are they? Marshall McLuhan, whose radical and prophetic book, *The Media is the Message*, predicted the all-pervasive impact of the Digital Age in 1967, wrote that "*Innumerable confusions and a profound feeling of despair invariably emerge in periods of great technological and cultural transitions*" adding that "*these are difficult times because we are witnessing a clash of cataclysmic proportions between two great technologies. We approach the new with the psychological and sensory responses to the old. This clash naturally occurs in transitional periods*"(McLuhan et al., 1967).

Indeed, transition is often accompanied by fear. Particularly important were transitions in the medium of communication. From story-telling, to written language, then the printing press, television and radio and finally, the internet. Take the printing press as an example. This ushered in an era where mass-produced literature could influence the many, not the few, allowing indoctrination and marketing to reach a hugely increased audience. Tracts, flyers and leaflets could be printed in their thousands, and distributed across the land. Martin Luther, the German theologist, exploited this to spread his doctrine at the start of the Reformation in Europe.

Pope Leo X proclaimed that the printing press had come down from heaven as a gift from God, but, later, concerns led to many printed works being banned by the Church. Certainly, the impact on European history, in terms of the printing press igniting the Reformation and spread of Protestantism from Martin Luther, is acknowledged. Sebastian Brant, the fifteenth Century German humanist, wrote that *"Through the genius and skill of the German people, there is now a great abundance of books"* (in Eisenstein, 2011). The power of the printed word, and the greater consistency, avoiding errors that could occur in hand-scripted texts, intentionally or otherwise, was seen as a real strength.

However, the printing press as a technological transformation was not celebrated by all. The Ottoman Turks banned the printing of any Islamic writings for three centuries, mostly because of opposition by the scribes rather than the religious leaders (Clogg, 1979). The famous Florentine book dealer, Vespasiano da Bisticci, denigrated the invention as *"made among the barbarians of some German city"* (in Eisenstein, 2011).

In 1499, Matthias Huss published a book entitled *La Grant Dance Macabre*, in which he depicted death attacking people involved in sinful activities. The book contains the first known image of a printing press, and the type setter, along with a book dealer and two others involved in the sinful business of printing, are being taken away by Death. Death tells the printers that there is no longer any time in which to correct their errors. Instead of errors creeping into individual versions of hand scripted manuscripts, errors in the printing process were replicated many times, and spread across a wide geographic area. This new technology could then be seen to multiply fake news by churning out many copies of error-strewn work, akin to fears relating to social media today. The new technology was also a threat to those in control.

Indeed, technological change can have huge impacts on society. The advent of mechanization on farms, the so-called green revolution, displaced some ninety percent of farm workers from their roles. The drive towards efficiency and reduced uncertainty, represented by the replacement of humans with machines, thus eradicating human error, has been a driver for much of the progress in manufacturing over the last one hundred and fifty years.

Just as with the printing press, individual errors are indeed eradicated by mechanization, and mechanical workers are less likely to have significant time off work for sickness or parenthood. But other dangers lurk. Errors can be spread virally in modern multimedia, effecting many systems, just like in printing.

The written word can be glorious or deadly, as can electronic media. In a recent event, the victory of the new media (information technology) over the old (the printed page) was demonstrated in an incredibly powerful way, as the disciple usurped its master. Two fans of the YouTuber, PewDiePie, hacked into thousands of printers that then printed out messages telling people to subscribe to PewDiePie's You Tube channel (Brewster, 2018).

When we get down to it, fears around AI seem to circle around a small number of issues. Firstly, AI may be so goal-oriented that it might eradicate us without even realizing. This is the classic *paperclip Armageddon scenario,* painted by Tim Bostrom (2003), a futurist based at the University of Oxford. Bostrom painted a picture of a certifiably innocent and harmless project but with shockingly dark consequences.

Bostrom describes the situation where a machine with AI is given a set of simple instructions: to make paperclips. Following along the lines of good practice, these are to be made as efficiently and cost effectively as possible. The situation seems straightforward: gather the raw materials needed to make paperclips (metal), then make them.

The AI recognizes efficiency in scale, and begins to build large factories to make the paperclips, while developing large mining operations globally in order to maximize the supply of metal. Soon, the planet is either one giant mine or one giant paperclip production line. When the mines run done, other sources of metal are exploited. Tin cans, cutlery, metal structures, cars, vans and lorries are all sequestered. Vast amounts of energy are consumed in the manufacturing process.

With the planet now turned into paperclips, there is no possibility of humans surviving. Indeed, since each adult human contains around four grams of iron (enough to make eight paperclips), mostly in haemoglobin and muscle, robotic vampires search the Earth, draining us dry of every last drop of the red stuff.

As Eliezer Yudkowsky puts it: "*The AI does not hate you, nor does it love you, but you are made out of atoms which it can use for something else*" (Yudkowsky 2008). The AI may also hide its activities from humans to prevent interference. The point is that it may be difficult for us to predict the outcomes. As Elizabeth Barnes, from Cambridge University, puts it: "*Humans are highly adapted to predicting the behaviour of other humans, but we may find it much more difficult to reliably predict the behaviour of AI systems*" (Barnes, 2016). While Paul Simon came up with fifty ways to leave his lover in his song of the same name, AI could well come up with millions of ways to do things, many of which we would have never thought of. Furthermore, we might not even realize it had left, until it was too late.

Alan Turing's (1950) idea of developing a much more basic '*child AI*', that we then nurture during its adolescence and parent through to a responsible adulthood, is riven with issues. To begin with, anyone who has been a parent will understand that you can't control anything, let alone everything, about how your child turns out. Adolescence sees a shift in the offspring, from relying on parents as mentors to relying on peers, and we have little or no input into the choice of peers or what they are like.

Traumatic events beyond our control can also hugely impact on the child, such as the death of a parent or experiencing an incidental trauma, such as a road accident. An AI, interconnected to the internet of things, will have multiple inputs beyond our cognition or control. There is the added complexity of balancing nature (the algorithms) with nurture (training data and interconnectivity).

Of course, the paperclip Armageddon scenario references another potentially existential risk, but one that is largely avoided by workers in the field: the catastrophic impact of AI on the environment. Sure, social and economic issues are important, but if AI continues down the path of using Nature as a sink and a source, as we have done for the last two centuries, but at a much higher efficiency and speed, then we are truly doomed.

If the goals set for narrow or general AI are the goals that we currently pursue, focused upon our own needs and desires, while ignorant of everything else, the consequences will be horrific. And this is a foundational point of this book. AI acting as an amplifier, extrapolating our flawed characteristics, will indeed create an apocryphal finale for humanity.

These types of fear are rooted in the failure of humans to properly educate their machines, and an inability to fully grasp the consequences of goal-oriented activities. Added to this is the issue of autonomy, something that no other technological advance has possessed. AGI could, at least theoretically, act without human input. Already today, the majority of the users of the internet are machines, not people (Lewis, 2016).

We can't just unplug an AGI, as it could well have access to its own power supplies, and, being networked to the internet-of-things, it would be impossible to bomb it into submission using shock-and-awe tactics without bombing the entire planet. Anyhow, the AI of the future could well control our very life support. How do you defeat an advanced cloud, with collective intelligence spread across the globe and, quite possibly beyond it, through the satellites that orbit around our world? It could probably take control of any weapons we hoped to turn on it. A true superintelligence would have all bases covered.

And there is plenty of evidence in our recent history to support concerns relating to our technological track record. Leaded petroleum, CFC refrigerants and diesel all serve as salutary lessons. Imagine if an AGI was tasked with restoring a sustainable planet. The obvious strategy may be to eradicate the greatest threat to this outcome, humans.

Another fear, long expressed, is that of an AGI becoming so superior in intelligence that it either deliberately enslaves us or destroys us. It maybe doesn't help that the word *robot* comes from the Czech word, *robota*, meaning *slavery*. Slave could become master. This idea of runaway intelligence (also called 'the fast take-off') with a dash of malevolence thrown in, is a common theme. The abuse of power has often appeared in human despots, so why not in robots?

Intelligence brings power and authority. In much the same way that we have lost respect for Nature, from which we came and upon which we rely for our survival, so machines, created by us, may one day lose respect for us. We may be viewed as the weakest link, and discarded.

A different set of concerns arrange themselves around the potential for AI systems to be deliberately designed to be malevolent. Roman Yampolskiy (2012) commented that "*If hazardous software is ever given the capabilities of truly artificially intelligent systems (e.g., artificially intelligent viruses), the consequences unquestionably would be disastrous. Such hazardous intelligent software would pose risks currently unseen in malware with subhuman intelligence*". But who would want to deliberately design malevolent AI systems? The answer is: lots of people.

Governments, opposition parties, military, insurgents, terrorists, criminals, businesses, doomsday cults or psychotic individuals could all have reasons to create destructive AI systems. Turchin (2015) gives some examples of how such malevolence could express itself. These include hacking other AI systems in order to steal their data and render them redundant, constructing robot armies to carry out aggressive activities through bioengineering or hacking large numbers of computers in order to gain more calculating power.

Self-drive cars could be hacked to kill people, or aircraft autopilots hijacked to crash planes into targets such as nuclear power stations. Hackers have already taken control of a self-drive car, interfering with its brakes, dashboard computer and door locks from twelve miles away (Solon, 2016), and of an aeroplane (Weise, 2015). Of course, genocide of the entire human race does not necessarily require a malicious evil genius. It is just as likely that humanity could be erased simply because it found itself coming between the AI and its goal. If the AI was extremely goal-focused, it may not even have to think about the decision, let alone target it as a sub-goal. Asimov's rulebook could be easily overlooked.

If AGI can be developed, what philosophical, ethical and political position would it take? What if the political philosophy of choice was anarchism? Would some form of religion or cult emerge, or some other cultural movement? What if this involved ritualistic human sacrifice? Another concern is the development of a form of moral nihilism, where the AI ceases to function because it cannot see the point in continuing. This philosophical landmine could lead to a now completely technology-dependent human race going extinct.

Then there's the problem of lightning-fast decision-making, something that computers are designed to do. As early as 1960, Norbert Wiener, the American mathematician and philosopher, reflected that: "*As machines learn, they may develop unforeseen strategies at rates that baffle their programmers. By the very slowness of our human actions, our effective control of machines may be nullified*" (Wiener, 1960). Within the context of wars controlled by AI, the concept of escalation may take on a whole new meaning. If the goal is victory in an AI-controlled engagement, a nuclear holocaust could happen, and within milliseconds. This is a truly doomsday scenario.

According to Dirk Helbing, each individual in the industrialized world has between three thousand and five thousand electronic records held within the big data, meaning that it is possible to map our actions and predict our decision-making directions. This in turn allows us to be nudged. *Nudging* is where tailor-made adverts and promotions are targeted at us, subliminally effecting our decisions, by gently pushing us in a particular direction. Like the little demon on the shoulder, these inputs gently whisper suggestions that seem to come from a voice inside us.

This brings the danger of homogenization. Homogenization is the process where we are all driven to think the same things, hold the same values and make the same decisions. It is the conversion of the human race to some Borg-like entity. For those non-Star Trek fans, the Borg was a fictional alien collective that assimilated other alien species, absorbing these species' knowledge and expertise, and turning the captured individuals into drones, to carry out the will of the collective. As the Borg famously say, "*Resistance is futile*".

Such homogenization, if taken to the extreme, could erase subjective experience and any form of debate, resulting in the zombification of the human race, where diversity and plurality are lost, as humans are unconsciously shepherded inside small conceptual cages. Such societal nudging, wherein all of us are gently helped in the same direction through a multitude of individual nudges, is a very real possibility.

This may seem oxymoronic, in that the internet surely opens up access to so much more knowledge than before? Yet search engines are controlled by algorithms, and so we are not often accessing the entire font of knowledge at all, but a subset that is pre-determined, either by the algorithm itself, or by businesses and other organizations who pay large amounts of money to promote their own sites over others. This means that we are all being fed what others want us to be fed, unless you are really determined to search out conflicting material.

We may imagine ourselves as brave explorers of the infosphere, but the mighty chargers upon which we imagine ourselves riding are really just packhorses, controlled by invisible forces and led to destinations that those forces want them to go to, with us on board, like Schopenhauer's cripple. The spectra of automated societies, where control, democracy and freedom-of-thought are taken by AI, and society becomes a product, shaped by the silicon gods, is a worrying one.

Of course, those in authority, be they governments or businesses, like homogeneity. If everyone can be trained to think the same and have the same knowledge resources, they are much more easily controlled and manipulated. Our education system targets homogenization, celebrating those who adhere to the rules best by awarding high grades in homogenous examination papers. Sure, the best students are rewarded, but the rest are taught to imitate and strive for the same standards as the best. The song, *Little Boxes*, by Malvina Reynolds, observed this acutely:

"And the people in the houses
All went to the university

Where they were put in boxes
And they came out all the same
And there's doctors and lawyers
And business executives
And they're all made out of ticky tacky
And they all look just the same."

My grandfather was educated in the early 1920s. He was naturally left-handed, but, at school, his left hand was tied behind his back and he was forced to use his right hand. This was another example of educational conformity. Diversity brings uncertainty to those in power, and that is never a good thing for the system that wishes to control.

Yet diversity is critical for the survival of any species or any ecosystem, in terms of resilience and multidimensional problem-solving. A good team is a diverse team. Being able to think in different ways and challenge the *status quo* should be seen as an immense strength. And so, these tendency towards converting 'developing' nations into 'developed' nations, of forcing the Enlightenment dogma, which has failed us so terribly, upon other people, and of denying that the planet is a pluriverse, resplendent and diverse, are strategies that are extremely damaging.

It is essential that AI does not catalyze such mass control, but, rather, that it promotes diversity. Because diversity and the process of diversification lie at the heart of any attempt at rebuilding and restoring a planetary system where we have a seat at the table. Quoting Helbing et al. (2016): "*Pluralism and participation are therefore not to be seen primarily as concessions to citizens, but as functional prerequisites for thriving, complex, modern societies.*"

Add to the mix the fear of robots taking over all of the jobs in the world, leaving us without purpose and with too much time on our hands, or the equally disturbing impact of robot-human romance and its impact on human-human relationships, and there seems to be a lot to worry about.

Fear surrounds AI. Thirty six percent of people surveyed by the British Science Association (2015) agreed that the development of AI posed an existentialist threat to humanity. Yet the majority of people out there probably don't have current knowledge of what AI actually is and how it works (unlike you, who, having purchased and are reading this book, and are now completely *au fait* with the field). Also, AGI doesn't actually exist at present, and so our worst-case scenarios are hypothetical.

This brings us to an important point. Much of the concern is coming from a variety of media, be it websites, movies or hearsay. And messages delivered by this media are largely constructed upon the two cornerstones that so often plague transitions: hyperbole and overstatement. Where do these questionable cornerstones come from? Generally, they come from two sources: academics competing for funding from government and private sectors, and businesses trying to impress customers.

Political parties can also contribute, as they seek re-election, promising utopian visions. And, of course, in order to use information as a weapon, whether it is to gain a grant, be seen to be doing good things or to gain further support, mass media is used, with messages often tailored to a particular audience, be they voters, shareholders or decision-makers.

Appealing to the general public through news bulletins released to websites, papers and broadcasters is often done, not particularly to convince the public, but to demonstrate to funders and management that you are active and promoting whatever organization you represent. And so, stripped-down soundbites, such as "*AI will deliver new cures for cancer*", "*AI will provide clean water for Africa*" and "*AI will make you rich and happy*" emerge. Or maybe the headline is "*robots could destroy the planet*", "*AI will exterminate the human race*" or "*AI will lock you outside of your spacecraft*". These banners help shift tickets at the local movie theatre, or help sell pay-for-view, satellite TV subscriptions.

Such exaggeration can stem from a number of sources. It may be accidental (built on true belief by the individual), deliberate, or misinterpreted (in that the person reading the story misinterprets it). It may be due to errors during retelling, wherein, over time, small changes occur, leading to a completely different message being delivered (as exemplified by the classic change of message where soldiers were asked to pass a message down the line, that said "*Send reinforcements! We are going to advance*", but, after a while, it had changed into "*Send three and four pence! We are going to a dance*").

Another source of error occurs when a particular report quotes one sentence out of context, and uses this to support a position never intended by the cited individual. A lack of historical context can also create issues, wherein an individual is quoted from a statement made in the 1950s, but who, with later developments, now holds a completely different viewpoint.

However, such hyperbole creates false hopes and fears. False hopes lead to increasing disillusionment and mistrust eventually, where discredence and cognitive dissonance develop. The Lighthill Report, in 1973, that led to the AI winter, was necessary in order to target exaggerated claims and promises about what AI could achieve.

Fears can destabilize, lead to stress and divide society, creating resistance, a lack of trust and uncertainty. There is a particular irony attached to the fact that the incredible and rapid increase in interconnectivity delivered by new technologies also brings with it the mass communication of hyperbole, catalyzing exaggeration and further hyperbole.

Fake news isn't new. Around 1258 BC, Ramesses II, Pharaoh of Egypt, had carvings made depicting him slaying the Hittites and conquering them, whereas, in reality, the conflict was viewed as a stalemate, resulting in a peace treaty between both sides (Weir, 2009). Ramesses painted one image to his audience at home, while acting in a completely different way with his negotiations with the Hittites. This sort of duplicity has been widespread through history. Imagine if Ramesses had access to the internet. And of course, privacy regulations may actually encourage such duplicity, wherein cloak-and-dagger activities can remain hidden from different parties through selective application of data protection legislation.

It is important to clearly state that many of the fears expressed are already becoming reality in narrow AI. In 2018, the journalist, Stephen Barnes, stated that *"There are only two types of companies: those that have been hacked and those that don't know they have been hacked"* (Barnes, 2018). Cybercrime is as widespread as the internet itself, and is estimated to be worth some six trillion dollars annually by 2021 (Morgan, 2017). Issues over privacy, nudging, political interference and commercial espionage are now regular news stories.

It is believed that flaws in risk-estimation algorithms, based on inaccurate modelling of default risk correlation (the *Gaussian copula*), led to the 2008 financial crash, whose ripples are still being felt more than a decade later across the globe (Salmon, 2012). Another example is Google's Flu Trends tool, that repeatedly misdiagnosed influenza trends across the USA (Lazer et al., 2014). Citron (2007) writes that algorithms are now *"the primary decision-makers in public policy"* rather than being merely aids to human managers. This being the case, flaws in these algorithms have huge potential to wreak havoc across society.

However, the fears related to superintelligence, in terms of our enslavement or annihilation, and those relating to all of our jobs being taken are far from being realized. As good as Watson is, it is not an autonomous, free-thinking agent with general intelligence and sentience yet. Currently, this is a very long way off.

Furthermore, the New Zealand philosopher, Nicolas Agar (2016), argues strongly that we are likely to solve the control issues around AI and emphasizes that *"Alarmist rhetoric is unhelpful. We should aim for a rational management of future technological risks that appropriately balances risks against potential benefits"*. Robert Brooks (2014b) makes a similar point, saying: *"I think it is a mistake to be worrying about us developing malevolent AI anytime in the next few hundred years. I think the worry stems from a fundamental error in not distinguishing the difference between the very real recent advances in a particular aspect of AI, and the enormity and complexity of building sentient volitional intelligence."*

John Danaher likens the work of futurists such as Bostrom to *sceptical theism*, wherein the existence of real evil would mean God could not exist, therefore, what appears to be evil actually is for the benefit of humankind, as a sort of lesson (Danaher, 2015).

This was exemplified by the character of Job in the Old Testament, who lost everything to the devil, including his children, but came out of it a better person, and got lots of things back. I've never felt good about Job. After all, if you were one of the kids, evil really did end everything for you, permanently, and you got nothing back. But Danaher argues that Bostrom's concept of the treacherous turn, wherein the AI pretends to be good but is really evil (the opposite of apparent evil being good) has significant implications, as it undermines our very ability to test for truth in any setting far beyond AI.

Thus, it undermines any sense of ethical value, leading to moral paralysis. Should all good be interpreted as evil? Does the treacherous turn mean there can be no benevolent forms of AI at all?

This doesn't mean we should not seriously examine the repercussions of such technology and how we design it, but we are still dealing with science fiction, not science fact nor even science extrapolation. However, the fears relating to superintelligence are important for a different reason. They shine the spotlight upon the threat that we as humans pose. The evil traits that we seek to detect in the reflective metallic eyes of a far-distant-yet-to-be-developed robot merely reflect the characteristics of our own species, and, if we look for long enough, we can recognize this.

Wreaking havoc upon our planet, our fellow species and ourselves, we are driven by an algorithm that represents greed, selfishness, autonomy and disconnectedness. This must change if we are to step away from the path we are inexorably charging along. And to help us turn off this path, we need to face up to our reality and what we are doing. But there is hope. We are not yet an automated society. Recent protests against climate change among the children of the world, led by the incredible Swedish school girl, Greta Thunberg, have reminded us that there is a deep awareness of what we are doing wrong and what we need to do right.

And so, we see that there are many issues with AI. However, these issues are mere extensions of humanity. We need a new approach, not an extrapolation of the flawed path we have beaten through the Earth. Vladimir Masch stated that *"The new paradigm of Superintelligence becomes preservation of mankind"* (Masch, 2017). This is failed thinking, repeating the mistakes of old, where everything, from Nature to technology, is used to keep us in place. This is what has gotten us in this mess in the first place.

It is now the time to examine what human intelligence actually is, and what other types of intelligence might turn up to the audition to be the framework upon which we should base AI.

SECTION IV The Thinker: Human Intelligence

"If something has never yet been done, it would be absurd and self-contradictory to expect to achieve it other than through means that have never yet been tried."

–Francis Bacon, *Novum Organum* (Bacon, 2010 [originally 1620] Book 1:6)

Welcome to section IV. We have now familiarized ourselves with the hardware and software of AI, the ethical challenges, inequality issues and the hyperbole and fear surrounding this new technology. Next, we need to turn our minds to human intelligence, vaunted, by ourselves (after millennia of reflection), as the one characteristic that sets us far from the madding crowd of bugs, bears, tumbleweed and trout. We begrudgingly admit that many organisms see better in the dark, run faster, pick up a scent better, cope with extremes in temperatures better and, basically, survive in the wild much more successfully than us. But we are the wise, wise ones, sentient, conscious and creative. For we are *Homo sapiens sapiens*.

So, what exactly are we attempting to artificialize with AI? We now examine this complex subject. It is as much to do with who we think we are as who we actually are. Navel-gazing does that. We discover that intelligence is a multi-faceted crystal. We explore some of these facets, including logical-mathematical, emotional, social and spatial intelligence. As we stare into this heart of darkness, we will tease apart what this says about our perceptions of intelligence and the implications for AI.

Philosophy is important here, and we take a crash course. Since our philosophical worldview greatly impacts on our thinking, which school of philosophy should we program into our AI? Could an AGI develop its own philosophy, and what could this mean for us? Imagine a Marxist AI working in a trading house. Would a nihilist AI even be bothered to do anything or would it just sit in its room like a moody teenager? It really makes you think...but, hey, we are the thinkers after all.

IV.1. Human intelligence: Carolus Linnaeus and his wise, wise men

Intelligence, be it artificial or real, is a very elusive and ethereal entity. All smoke and mirrors, just when you think you've pinned it down, it slips through your fingers, like the sand in that desert where the thirsty frog lived. Akin to Plato's shadows on the cave wall, the further you look, the more complex and difficult the concept becomes. Yet we have to endeavour to grapple with this shape-shifting concept if we are to deconstruct AI. It also lies at the heart of the greater issue here, how we found ourselves in our present plight, and what we need to do to escape. There must be some way out of here, said the joker to the thief, but to find it we need to understand the concept of intelligence. Our search for who we are, where we come from and where we are going lies at the heart of this book.

Carolus Linnaeus, the Swedish taxonomist, was a man who believed that revenge was a dish best served cold. Johann Siegesbeck, the German botanist, had referred to Linnaeus' work as loathsome harlotry, possibly referring to the Swede's description of the group of plants called *Enneandria,* as nine men in the same bride's chamber. It was the 1700s, and Linnaeus' attempt to make his new taxonomy popular, by caricaturing these characteristic anatomical properties in a somewhat bawdy, music hall style, didn't go down well at all, and not just with Siegesbeck. However, Linnaeus remembered Siegesbeck's slur, and a few years later, he immortalized the German, using the weapon he possessed, taxonomy, by naming a small, unpleasantly sticky, insignificant weed after him: *Siegesbeckia orientalis*. It is unclear if Linnaeus was aware of this plant's medicinal use in the treatment of syphilis.

On another occasion, following a disagreement with one of his students, Daniel Rolander, Linnaeus named a species of beetle as *Aphanusrolandri,* meaning *inconspicuous Rolander*. Nor would this be the last time taxonomy was used in an underhand way. *Dermophis donaldtrumpi* comes to mind, a newly discovered blind and burrowing amphibian from Panama. On the opposing political aisle, a taxonomist with a dislike for former Soviet leader, Nikita Khrushchev, named a worm genus, *Khrushchevia,* after him, possibly in reference to his infamous line, aimed at Western nations, *"We will bury you"*.

Staff at the Carnegie Museum of Natural History in Pittsburgh were known to dislike their director, William Holland, who, they felt, hogged all the glory of their research. In due course, they named an extinct pig species by the binomial of *Dinohyus hollandii,* literally, 'terrible pig Holland'. One of my favourites, and not at all insulting, is the Costa Rican parasitic wasp, named *Heerzlukenatcha,* and, to finish, the northern Venezuelan fly, *Piezaderesistans*.

Of course, all of this points to a rather more depressing nuance: our treatment of Nature as our plaything. Not that the plants and animals of the world care less about what we call them, but it reinforces, subliminally at least, our sense of control over Nature, and, somehow a lack of respect and a failure to recognize its importance and sovereignty.

Linnaeus is best known as the inventor of the *binomial system* for naming species. Each species was given a two-name (binomial) title. The first was the *genus,* the second the *species*. It works a bit like a noun and an adjective. If we had the noun *ball,* we could have different adjectives such as *big, small, red* or *yellow*, defining what kind of ball it was. Species are treated in the same way. Each organism is assigned to a genus (akin to a noun) and then a second name is added (akin to the adjective), the species name, that separates the particular genus into more specific groups. Only organisms in that particular species share these more specific characteristics.

Linnaeus didn't actually invent the binomial system at all. Many botanists before him, such as Konrad Gesner, Pierre Richer Belleral and John Ray, had already been using this approach years earlier. However, it was Linnaeus who set out the rules to universalize the approach.

And it was Linnaeus, in 1758, who gave our own species its binomial. In the first volume of his book, *Systema Naturae*, he named us *Homo sapiens*, literally, *wise man*. Eventually, as other sub-species were discovered, we'd be elevated to our own sub-species, *Homo sapiens sapiens*, or *wise, wise man*, in case someone was stupid enough to have missed it. You can't get much wiser than that: wise squared.

Indeed this high regard for ourselves, the only species called *sapiens*, stretches back in time in literature. The creation story shared by Islam, Judaism and Christianity, proclaims that humans were created in the image of God and would rule over Nature. The biologist Ernst Haeckel, in his effort to construct a tree of life, charting the relatedness of all living things, put humans at the top of the tree. From this elevated haughty roost, we gazed down on the rest of the biosphere.

Marie-Jean-Antoine-Nicolas de Caritat, Marquis de Condorcet, the great French revolutionary and educationist would write, "*Nature has fixed no limits to our hopes*" (Condorcet, 1955 [originally 1779]). We may have been the product of its tormented and mindless struggle, but we were better than this. And what stood us apart from the tooth and claw of Nature, was our intelligence. Émile Durkheim, the sociologist, claimed that "*it is civilization that has made man what he is: it is what distinguishes him from the animal: man is man only because he is civilized*" (Durkheim, 1973 [originally 1914]).

This concept of civilization is of interest, partly because it appeals to an artificial construct, rather than the natural, biological construct. Here, civilization can be viewed as beyond the gene, or extra-chromosomal. Every living thing has genes, but civilization is not a property of all, according to Durkheim.

This leads to another question: is intelligence an outcome of the natural (i.e. genetic), of the artificial (i.e. civilization) or of both? The reason this is so interesting is because it brings us to a core consideration of this book. What is artificial intelligence, and how do we compare its reality and potential to our own intelligence? Ultimately, an examination of this issue returns to roost at home. For we then have to ask what we, ourselves, are and what guides our decision-making.

Tennyson reflected on Nature as raw in tooth and claw, and Darwin wondered at how such perfection could emerge from such cruelty, writing: "*Thus, from the war of Nature, from famine and death, the most exalted object which we are capable of conceiving, namely, the production of the higher animals, directly follows*" (Darwin, 1994 [originally 1859]). And the most exalted of all the exalted is the wise, wise one, *Homo sapiens sapiens*.

We have even developed our own politico-philosophical school, *Humanism*, celebrated by Enlightenment philosophers, economists and revolutionaries since the eighteenth century. Humanism dominates Western thinking and more recent development theory. Humans, through reasoning and technology, can beat a path of progress to their own utopian futures.

And while God may not have created us, we have become like gods, by creating an intelligent being of our own. Not some piecemeal construction of previously existent body parts, scraped together from bits of other bodies as in Mary Shelley's vision, but a humanist creation tale all of our own: then humans said, "*Let us make an artificial intelligence in our image, in our likeness*".

For, surely, human intelligence is the only truly meaningful intelligence? Crows can use sticks to gain rewards. Monkeys have a few tricks up their sleeves. Otters smash shellfish on stones. But we don't really view any of these as intelligent, sentient beings. They've just stumbled across a few tricks. Artificial intelligence should mimic human intelligence, not the mayhem of the natural world, where selfish genes direct the chaos and the killing. It's a no brainer. Or is it?

IV.2. So what is human intelligence?

Human intelligence as a subject of study has, itself, pre-occupied the human intellect for thousands of years. In the ultimate self-indulgent exercise, we have long hypothesized over what makes us tick, how we think and what this says about us. How to understand and control ourselves or others, what motivates us, how we make decisions, solve problems and imagine our futures are all things that have consumed the time of philosophers, priests, psychologists, psychiatrists, politicians and the rest of us probably since humans first thought.

Religions have thrived on such ponderings, as we contemplate our relationships with the cosmos and each other. Self-help books, counsellors and gurus have exploded upon the human landscape and our thirst for answers has become a multi-billion-dollar industry. Reflection and mindfulness are the new norms, as we search for an understanding of our psyche. We have each found our own sages, from Freud to Jung, and from Dr Ruth to the Maharishi Mahesh Yogi.

But what is human intelligence? Legg and Hutter (2007) reviewed the literature and found at least seventy different definitions of intelligence. Working through these definitions, they identify three common themes: firstly, the centrality of interaction with the environment; secondly, success in achieving an objective; and thirdly, flexibility in response to different challenges. From these, they define intelligence as a measure of "*an agent's ability to achieve goals in a wide range of environments*" (Legg and Hutter, 2007).

Edwin Boring, the Harvard psychologist, was more scathing, when he wrote *"Intelligence is what is measured by intelligence tests"* (Boring, 1923). In other words, it is so vaguely understood as to be subjective rather than objective. Indeed, the very testing of intelligence, based around intelligence quotient (IQ) assessment, is hugely questionable. Depending on the type of test we use, we can get very different measures of how intelligent a person is. People think in different ways.

IQ tests are a contentious aspect in intelligence theory and practice (Mackintosh, 1998; Kaufman, 2018). There are a number of areas where problems arise here. Firstly, there is the formal examination issue. Any written test is targeted at a narrow subset of what we think of as intelligence. I know many brilliant individuals who failed miserably in school examinations and SAT and GRE tests. One of the best cricket players I ever encountered, who captained our school team to relative glory and was a tactician beyond tacticians, never scored more than twenty percent in his formal mathematics examinations.

For many others in the past, learning 'conditions' such as dyslexia prevented them from doing well. Sir Jackie Stewart, three-time World Drivers Champion, leading business man, and dyslexic, famously was told by his teacher when he was twelve years old *"Jackie Stewart, you're stupid, dumb and thick"* (Clash, 2016). Yet IQ tests are serious business. If a prisoner is sentenced to death in the United States but has an IQ score lower than seventy, he will not be executed (Kelly and Resnick, 2014).

Individuals with large long-term storage and retrieval capabilities (often dubbed 'photographic memory') can sail through formal school and university examinations. However, state examinations and IQ tests generally fail to encompass other aspects of ability such as social, artistic, emotional and ecological skills, that are essential aspects in terms of functional intelligence in the real world.

This may be why my genius cricket captain failed so abysmally to succeed in academia. In other words, IQ tests and other assessments of intelligence are biased towards particular ways of thinking. This bias is state-sponsored, as government exams lead the way here. The state then ensures that individuals with particular types of intelligence succeed and become leaders of society, reinforcing the "norm", as dictated by the state. As globalization of education standards spreads, this becomes a global norm.

Situated intelligence is the theory that sets out the idea of intelligence emerging from the close interaction between the individual and their environment, wherein any understanding of our intelligence must rest on the fact that we are situated within that environment. This resembles *Gestalt theory*, that claims that the conscious experience must be considered as a totality, i.e. that consciousness embraces the entire system, not just the conscious individual. Thus, society and the natural environment are part of this totality.

Over time, many efforts have been made to distinguish different forms of intelligence. This is a difficult area. Are we just observing the same thing, say, general intelligence, from different angles or truly separate functions? For example, David Wechsler, the Romanian psychologist, stated "*social intelligence is just general intelligence applied to social situations*" (Wechsler, 1958). Could the same be said for all of the other categories?

Emotional intelligence is defined as the "*ability to monitor one's own and others' feelings and emotions, to discriminate among them and to use this information to guide one's thinking and actions*" (Salovey and Mayer, 1990). It encompasses empathy, relationship building and forgiveness, while providing a sense of who we are in a particular context. Emotional intelligence appears as a glue that sticks society together, allowing individuals to influence and be influenced.

Social intelligence, while sometimes argued to be merely an outcome of emotional intelligence, has itself been a major field of research since John Dewey (1909). Social intelligence has proven extremely difficult to measure, unlike IQ. Frank Landy, the American psychologist, characterized the search for social intelligence as "*long, frustrating, and fruitless*" (Landy, 2006). Edward Thorndike defined social intelligence as "*the ability ...to act wisely in human relations*"(Thorndike, 1920). The concept has developed, now embracing political as well as emotional intelligence. Leon Bosman (2003) writes that social intelligence is a broader construct that subsumes emotional intelligence. Marlowe's (1986) model of social intelligence comprised five domains: confidence, empathetic ability, pro-social attitude, emotional expressiveness and social performance skills.

George Herbert Mead, psychologist and founder of the Chicago Sociological Tradition, where the mind was viewed as emerging from social interaction, stressed that "*It is absurd to look at the mind simply from the standpoint of the individual human organism. We must regard the mind as arising from the social process*" (Mead, 1934).

How broad is this concept of sociality? Is it restricted to humans, or can our interactions with other species, landscapes and ecosystems also contribute to the emergence of the mind? What of artificially intelligent systems? Alan Wolfe, in 1991, asked "*Can communication between a human and a machine be considered mindful?*". Since the social self requires other selves in order to exist at all, so too no human can possess reflective intelligence (i.e. a mind) without other minds in other individuals also being present.

And in a way, unless the AI was human, it could not possess a human mind, as the mind would have to be human in order to interact with another human mind. This has become of increasing interest in light of the introduction of autonomous social robots as carers for dependent people, particularly in retirement homes (Draper and Sorell, 2017). Concerns exist that their lack of social and emotional intelligence could disturb humans, particularly if they are on the dementia spectrum.

However, given that the internet of things, with its seven billion eyes and ears, has the ability to sense vast amounts of social and environmental data, AI can be situated within this broader context and has the potential to develop a form of Gestalt thinking, wherein it embraces the entirety of its context in a holistic way. We will see that this is crucially important in converting this technological wonder into a transformational tool for the human race.

Diametrically opposed to situated intelligence theory is the theory of *successful intelligence*. This posits that "*intelligence is defined in terms of the ability to achieve success in life in terms of one's personal standards, within one's sociocultural context*" (Sternberg, 1997). Very much focused on the individual as the unit of intelligence, this approach fits the western philosophical position, as opposed to a greater emphasis on social intelligence as seen in African, Andean and Asian thinking, wherein the individual is defined within a societal or broader environmental context.

The traditional Enlightenment vision of AI is to bring progress and success, particularly in an economic context, wherein technology delivers a trickle-down effect. Here, wealth replaces social and environmental capital. By generating greater efficiency in production, greater profitability and greater GDP through exponential economic growth, it is argued that everyone will benefit, and the perfect society can be bought and paid for, made up of wealthy, healthy, educated and, thus, successful individuals. The theory of successful intelligence reinforces this vision of AI.

Furthermore, at the heart of this theory is the concept of *shaping*. This involves the ability to modify the environment to suit oneself. This reflects the lack of linkage with the environment, whose role is merely to be utilized and shaped, the classic sink/source relationship between humans and Nature. This reductionist approach stands in sharp contrast to that of systems theory, where shaping is an emergent property that impacts on all constituents of that system, rather than a particular constituent (e.g. the human) altering everything for their own optimality.

Multiple Intelligence theory (MI theory), developed by the Harvard psychologist, Howard Gardner, sets out eight different types of intelligence: linguistic, logical-mathematical, spatial, bodily/kinesthetic, musical, interpersonal, intrapersonal, and naturalistic (Gardner, 1983). Interpersonal and intrapersonal intelligence have also been considered together as emotional intelligence. An individual who has a particular strength in one type of intelligence will not necessarily demonstrate a comparable aptitude in another intelligence. Thus, from Gardner's perspective, traditional IQ tests are not adequate in that they generally only test one or two of the eight types of intelligence.

MI theory asserts that formal educational settings (i.e. schools and colleges) emphasise only linguistic and logical-mathematical intelligence, largely ignoring the other six forms, and this is a self-perpetuating process. In terms of AI, how could multiple intelligences actually operate? Would there be a blending somehow, or a system with multiple personalities, rapidly switching between, say, logical-mathematical and naturalistic intelligence as required? This could be extremely challenging. Indeed, with eight or possibly more modes of intelligence, trade-offs would be required. How could this work? What kind of algorithm could encompass all eight forms of intelligence? Of course, the logical-mathematical intelligence is clearly straightforward, but what of intrapersonal and interpersonal intelligence?

Another problem is that MI theory suggests that each person has a different combination of the eight intelligences. How then would we decide which combination a given AI would have? MI theory claims that there is no co-ordinating executive function that decides what combination of intelligences should be combined for a particular problem. Instead, it suggests that a decentralized model of organization operates. MI theory, like so many studies in intelligence, leaves many questions as to its functional and neurobiological mechanisms within humans.

This comes back to a recurrent problem with basing AI on human intelligence. How can we hope to design AI in our own image if we can't really see what our own intelligence looks like? The observer effect also impacts here, as the process of navel-gazing is still observation, thus potentially altering our perception of ourselves.

Finally, if MI theory holds true, differences within a population, in terms of the combinations of these intelligences within each individual, will contribute to diversity, and thus resilience. Surely then we need to be attempting to develop such variation across a population of AI units. Maintaining such diversity, when data impacts on the functioning of, in particular, non-symbolic AI systems, may be difficult. The variations in intelligence may need to be hotwired in a more symbolic way.

As much as we recognize that the technology still does not exist for artificial general intelligence (AGI), we need to recognize that our understanding of human intelligence is equally undercooked. Of course, the one advantage that we have in our study of human intelligence is that we are humans ourselves. Therefore, while it is a bit like trying to lift ourselves up by our own shoelaces, at least it is better than trying to understand cat intelligence, where we will never be cats and so have really very little idea of how a cat thinks.

Thus, although constrained by what we think we are, we can still have the potential to make some progress. However, to decide that human intelligence makes the best model upon which to base artificial intelligence is another world entirely. Not only do we not have a consensus on how our minds work, but the evidence of the failures of our thinking are so clearly apparent, as is the diversity of outputs. The same basic brain neurobiology that equipped Nelson Mandela and Martin Luther King with the ability to transform their nations also equipped Hitler, Pol Pot and Stalin with their own powers of dark persuasion.

Dreyfus, in 1965, wrote "*Is an exhaustive analysis of human intelligent behaviour into discrete and determinate operations possible? Is an approximate analysis of human intelligence, in such digital terms, probable? The answer to both these questions seems to be, No*" (Dreyfus, 1965).

IV.3. Philosophy and intelligence: The framing of our thoughts

Another issue arises in terms of philosophy. Even if we decide upon a theory of human intelligence, which school of philosophy should our AI adopt? Each of us practices philosophy, whether we like to admit it or not. For example, a belief in science requires *empiricism* (the theory that all knowledge is based on experience derived from the senses). This greatly shapes our understanding of the world around us, where we follow the path of reductionism and pursue cause-and-effect reasoning. Much of the Western world is shaped by empiricism.

However, we could pursue a very different philosophical path. Take *existentialism* for example. Existentialism argues that the world is not an ordered, determined system understood through the construction of laws based on observation. Within this philosophy, worldly pursuits are viewed as futile, and personal responsibility is key. Society is viewed as an artefact, and often a negative force. The individual should determine their own journey, unfettered by society, state or religion. An existentialist AI would, potentially, be on a collision course with a capitalist worldview.

Relativism emphasizes that context is everything. Here, *situation ethics* rules, wherein it depends on the context if something is right or wrong. This very much ties in with situated intelligence. For example, it would be wrong to kill unless by killing you saved more lives. Imagine if an AI developed relativism as its philosophy. It might well ignore the four laws of Asimov (robots shouldn't hurt humans, robots should obey humans, robots should protect themselves and robots should protect humanity) in given situations.

Or what if AI adopted *Marxism*? By eschewing fundamentalist capitalist values, and seeking to redress the reduction of everything to a commodity, this could really prove interesting in an AI-controlled financial market. A *Nihilist* robot could be even more fascinating, although it might just sit in the corner and feel that no problem was worth wasting the energy on.

Philosophy informs decision-making and ethics and is a central component of intelligence, colouring and flavouring our interpretation of the world, and creating a representation of how we understand self and others. It contributes to so many aspects of our thinking, from emotional intelligence through to ethical and moral judgements. If strong AI truly emerges, then there is every possibility that artificial life will explore the world of philosophy, and identify with a particular variant, impacting hugely on its functioning and development.

Even weak AI, reliant upon symbolism, will need some sort of philosophical framework if it is to function within the grey areas of life, where more complex issues arise. This will depend on the programmer, but will have, potentially, huge impacts in terms of decision-making, trade-offs and problem-solving. There is the real danger of philosophical bias from the white, male, Western dominance in programming. The most likely prevailing approach is Western empiricism. Such philosophical bias could be seen to represent another example of neo-colonialist supremacy, joining sustainable development, neo-liberalism and trade-for-aid-and-ammunition (i.e. we loan you money and provide military power and you let us control your oil reserves) in the re-enactment of empire.

The philosophical foundations of any thinking organism are essential elements in their functioning. Decision-making forms the bridge between ideology and action, but philosophy determines ideology. Ethics, motivation and the representation of the world beyond the individual are seen through a philosophical lens. Yet this whole aspect of intelligence has been ignored in the AI literature. And that is a significant issue.

We can see that human intelligence is a disputed field in academia, while in real life, it has resulted in the decimation of the Earth system. Of course, we are also capable of beautiful things. We have designed and built incredible structures, such as the pyramids of Giza, Angor Wat, Machu Picchu and the Basilica of San Marco. From van Gogh to van Beethoven, our creative genius has reached extraordinary heights. Indigenous races live in harmony with their environments. Yet there are serious flaws in how many of us live our lives.

As we begin to shape the intelligence at the heart of AI, a technology that has the potential to do great good or bad, is the human model really the best approach? Let us now consider what other types of intelligence are out there, and what they can bring to the table in terms of AI.

SECTION

V

Other Modes of Intelligence: Thinking Outside the Human Box

"In the past, the man has been first. In the future, the system must be first."

–Frederick Winslow Taylor, *Principles of Scientific Management* (Taylor, 1911)

All life forms face problems and challenges, and must resolve these in order to survive. In multicellular organisms, this is a co-ordinated effort. Of course, single-celled organisms can also co-ordinate through living as part of a colony. Indeed, many organisms find their fate intertwined, within ecosystems. Communication is key, usually through chemical signalling. Yet we still tend towards the intuitive, if erroneous idea that only organisms with brains are intelligent, and, indeed, that *Homo sapiens sapiens* is especially and uniquely intelligent.

Hence, artificial intelligence is based, unashamedly, on human intelligence. But in this section, we challenge this view fundamentally, suggesting that in many ways we are the least successful model to base such potentially significant technology upon. We now examine intelligence in animals, plants and bacteria, discovering that they have very different ways of thinking than we do. Then we turn to ecosystem intelligence, the cathedral of cognition where the diverse congregation of thinkers gathers together. Here we discover systems theory and begin to grapple with what it means to be a system. This is the most important part of this book, so strap yourself in.

V.1. Animal intelligence: Machiavellian sentience and the wisdom of the swarm

Next down the pecking order from humans, on a lower set of branches on Haeckel's tree of life, roost the animals. Raw in tooth and claw they may appear to be, but they live within their environments in a much less destructive way than we do. Some share similarities with ourselves, especially the vertebrates. Having a backbone somehow endears the other vertebrate species to us.

We are much more likely to have a pet vertebrate than a pet invertebrate. And we even begrudgingly admit that some members of the Animal Kingdom (more correctly, *Kingdom Animalia*) display some form of intelligence.

However, our attitude is generally fairly condescending to our animal neighbours, unless we perceive them as a threat to our physical safety. Dr Doolittle may have talked with the animals but, really, how interesting were those conversations?

Rodney Brooks has argued that rather than aiming to mimic human intelligence directly, we should build our way towards the human by starting with simple creatures as models. He suggests that the typical house fly, for example, is unlikely to be *"recovering three dimensional surface descriptions of all the objects within its field of view, reasoning about threats from a human poised with a fly swatter, in particular about the human's goal structures, intents or beliefs, representing prototypes and instances of humans (or coffee pots, or windows, or napkins), making analogies concerning suitability for egg laying between dead pigs and other dead four legged animals, or constructing naive physics theories of how to land on the ceiling"* (Brooks, 2014a).

Herbert Simon (1969), in his classic work on artificial intelligence, *The Sciences of the Artificial*, uses the example of an ant making its way across a beach. If we plotted its path, it would change irregularly, as the ant avoided a rockpool, or crawled around a stone. It might cut across a pheromone trail or encounter a predator. It might suddenly come across a wave breaking on the shore, leading to it being washed higher up the beach or out to sea. In all of these situations, the complexity of the path taken relates to the ant's environment and the apparent behavioural intricacy may not represent a complex inner system in terms of the ant's cognition.

Thus, Simon argues, we may wrongly interpret behaviour by thinking of it as autonomous and not environment-dependent. This is something we are really good at. When cane toads were released in Queensland, Australia, no thought was given to the very different environment of Queensland compared to Central America, and the repercussions of that upon the behaviour of the cane toad.

Meanwhile, Agre and Chapman (1987) went further, suggesting humans don't tend to operate at the sophisticated levels at which we are attempting to design AI either, but that our lives are mostly concerned with carrying out fairly basic tasks. This resonates with Morgan's Canon. This states that we should never invoke a higher order explanation for behaviour when a simpler one will suffice (Morgan 1894, but see Thomas (1998) for an insightful discussion on the misuse of this canon).

However, it would also be wrong to assume that animals are merely flotsam and jetsam, moved by the environmental tides. There is ample evidence that many animals have extremely advanced cognition and intelligence, rather than using it as a stepping stone, as Brooks posited. Given their integration within the Earth system, Wilson (1991) suggests that we should create AI by modelling animal intelligence, pure and simple. So, what kinds of intelligence do animals exhibit?

Fish have been shown to possess consciousness, Machiavellian intelligence, long-term memories and complex traditions (Brown et al., 2011).

Fish can recognize each other and prefer to shoal with individuals they know. Shoals of fish who know each other are more likely to avoid predation than shoals made up of strangers (Chivers et al., 1995), a good example of social intelligence, wherein shared familiarity and knowledge increase survival. Fish such as wrasse also make use of tools (Bernardi, 2012). Other research suggests that invertebrates such as crayfish display sentience (Fossat et al., 2014).

Of course, multicellular animals have a circulatory system and a nervous system, key components in terms of communication and behaviour. These systems deliver what the cognitive centre decides, and make such creatures much more familiar to us in terms of structure. We can relate, at least to the underlying wiring that permeates them and us.

Animal behaviour has become an important model for AI, to such an extent that the term *swarm intelligence* (the collective behaviour of a group of entities) is more likely to refer to algorithms used in cellular robotic systems, wherein a number of AI units interact together, than to animals such as the social insects, where it was first studied. These algorithms have been developed from numerous different animal models, including bats, monkeys, wolves, glow-worms, bees, ants, fireflies and krill (Kumar et al., 2016; Chakraborty and Kar, 2017).

Fundamentally, social interaction facilitates information sharing, and information is power. Whether it is reading the complex messages of a pheromone trail or alerting others to the proximity of a predator, information is important. Animal behaviour has at its heart the transfer of information, something that also lies at the heart of information technology and AI.

Swarm intelligence is found in many different species. First discussed in a general way by the French revolutionary, the Marquis of Condorcet, in the eighteenth century, he demonstrated that the larger a group was, the more likely that the majority would make the correct decision (Sumpter and Pratt, 2009). However, Condorcet assumed that each individual will make their decision independently, but this is not necessarily always correct. Peer pressure, bias and relatedness can impact on independence. A clever but flawed politician can convince a crowd to follow him even if he is fundamentally wrong.

Bees reporting on the benefits of new nest sites will dance a *waggle dance*, indicating where the new site is located. The longer the dance goes on, the better the site. Other bees who are impressed with this communication will also visit the site, and, if they like it, they will return and start dancing about it too. This, in time, leads to a comprehensive vote for the best site. A group of related individuals may be genetically prone to a particular action, or may participate in kin selection, wherein social interactions, including group decision-making, are suggested to be correlated to genetic relatedness. J.B.S. Haldane put it this way *"Would I lay down my life to save my brother? No, but I would to save two brothers or eight cousins"* (Haldane, 1926). It should be noted that there has been much criticism of kin selection (see Kramer and Meunier, 2016).

Importantly, a differentiation must be made between organisms making their own decisions within a social context, and the situation, defined as swarm intelligence, where individuals gain information independently but then combine this information to create a solution, overcoming each individual's cognitive limitations. James Surowiecki (2004) refers to this as the wisdom of crowds.

Swarm intelligence is particularly recognized in social insects such as bees, termites and ants. Leaf-cutting ants have complex agricultural societies, growing crops of fungi on mulch made from leaves that they cut and bring into the underground fungal gardens. Termites build huge, elaborate nests complete with air conditioning that can be four thousand years old and ten feet tall. Honey bee swarms have dynamic task assignment and express collaborative intelligence. Swarms have no direct centralized supervision yet the hive of around twenty thousand individuals operates in a highly co-ordinated manner. Swarm intelligence has been closely studied and modelled in AI systems. The attractions are obvious. Relatively simple individuals can interact and communicate in such a way as to produce extremely complex outcomes, and can solve problems by self-organization in order to survive in a sustainable way through time. This society works as a unit, cohesively bound by lots of tiny communicative moments.

Many social insects such as ants use a process of communication called *stigmergy*. Stigmergy is a mechanism of indirect coordination, wherein the trace left by an action in a medium stimulates subsequent actions. This can lead to complex, co-ordinated group behaviour emerging without the need for planning or control (Heylighen, 2016). For example, a trail of pheromones may be laid down. Other members of the group can detect these chemical cues and indirect co-ordination follows, wherein the trace left by one individual stimulates subsequent actions. What results is a complex activity without any direct communication. Self-organization emerges, with the chemical signalling providing positive and negative feedback, as other individuals contribute to the pheromone trail, providing a quantitative measure of how popular the route is.

Finally, the reductionist approach of mimicking animal behaviour, movement and sensory aspects is a strategy of doubtful value. By plucking aspects of an animal's behaviour out, modelling it in a series of algorithms and putting these to work, as has been done time and time again, the context of the organism is completely ignored. This approach is what is termed *biomimicry*, and has become widespread throughout the design world. However, biomimicry is centrally flawed as without the appropriate feedback, context and spatio-temporal environment, both living and inert, what is mimicked would not actually exist in the real world. This is because life is set within socio-environmental contexts, and only makes sense within these contexts.

You cannot pick-and-mix, taking bits from here and there and designing some sort of Frankenstein-like intelligence. The other point to make is that life is arranged in a series of levels of organization, from atoms, molecules, cellular and multicellular individuals, through to populations, colonies and ecosystems. Each of these levels has its own emergent properties while each is informed by the other levels.

As a whole, the complex totality of life, which we refer to as the Earth system, represents a system with emergent properties and is a self-organizing entity. It is not a reductionist structure, a construction of small building blocks, where you can just swap blocks from a big box of shapes. Rather life is a much more dynamic, integrated, emergent entity, sustaining itself through time yet changing constantly. This is the vision we should embrace. AI must be context-based and fully integrated within the biosphere if it is to deliver a sustainable future. We shouldn't be practicing biomimicry, but, rather, *bio-participation*.

V.2. Plant intelligence: Headless, brainless, dispersed intelligence

Plants and algae are the gateways of energy into the biosphere. Through photosynthesis, they represent the batteries of the planet, powering life and allowing complexity to exist in an increasingly disordered universe. As a bridge between the Sun and the living world, their significance is unparalleled. But are they intelligent? Vegetarians and vegans feel fine eating vegetables but not meat. No-one feels they have killed a plant if they pick a flower in the meadow. Plants have no brains or heads. You surely can't be intelligent when you don't even have a brain? Plants usually can't move, although their seeds and pollen can travel huge distances. They just sit there, easy targets for a passing herbivore.

However, this conception of immobility is incorrect and misleading. In addition to their pollen and seeds, plants continuously explore their environment both through growth and chemistry. Roots and shoots explore the soil and aerial environment respectively, while roots pump chemicals into the soil to free up nutrients. Many plants, such as strawberries, can spread by runners, producing identical plants that then become independent. Other plants, such as ivy, can grow huge distances. The largest organism in the world is a plant that lives in Utah. It is a male Aspen, which has many stems but one root system, and it is nicknamed Pando. Pando weighs thirteen million pounds (six million kilogrammes) and covers an area of one hundred and ten acres (44 hectares).

Many plant parts can move, such as the traps of the Venus fly trap (a meat-eating plant), the leaves of *Mimosa* (that fold up if a herbivore is sensed) or the leaves of wood sorrel (that close to protect themselves from high levels of sunlight). Plants can bend and change direction in response to wind, touch, moisture, temperature or gravity. Plants also make structural changes in response to parasites, predators, geomagnetic fields, oxygen, electric fluxes, excessive light, chemicals and fluctuations in water flow.

Plants can live a lot longer than animals. Bristlecone pines, for example, can live for up to five thousand years. The giant saguaro cactus is from the Sonoran Desert, living for up to two hundred years. While it grows up to sixty feet tall, after ten years it is a mere one and a half inches in height. They don't grow their first arm until they are at least seventy-five years old. They develop within a completely different time scale than humans. This is important because it provides a key insight – plants aren't like us. What I mean by that is that we can't compare them to us, and can't interpret their intelligence through an anthropocentric lens.

In fact, most plants aren't plants at all. Ninety eight percent of plants are actually dependent on fungi, that grow within their roots and, often, inside their cells, for their survival. The fungi connect plants to each other, allowing them to share signals and resources. In fact, these plants are actually plant-fungi, called *mycorrhizas* (literally *fungus roots*). We now know that the fungi only link certain species of plants together, giving them an advantage over other plants. The fungi actually have a significant role in determining what species of plants can be part of a given community. They are social engineers.

Plants are considered to be more adaptable than animals and have much more flexible developmental pathways (Bowler and Chua, 1994). Since they can't move *per se*, they need to use these pathways to confront the issues facing them. Plants display intentionality, with above ground parts growing towards the Sun (and often tracking it across the sky) while roots explore the soil, searching for nutrient-rich patches and then proliferating when they find one. It is estimated that some thirty percent of a plant's genome is associated with root development (Zobel, 1975).

Plants can discriminate and differentiate between apparently similar things. They demonstrate self-recognition, wherein if one of their own roots encounters another of their own roots, they do not react in the same way as when they encounter a different plant's roots, even if that plant is a genetic clone of itself (Gruntman and Novoplansky, 2004). So, it is not genetic difference that is being determined, but, rather, a true sense of individuality. Plants will produce chemicals to poison a herbivore when they are attacked, but will not produce these chemicals if the damage is mechanical. One such anti-insect toxin is *delta*-9-tetrahydrocannabinol, the psycho-active ingredient of cannabis (Rothschild et al., 1977). Plants previously attacked by a herbivore will also elicit this response much more quickly the next time they are attacked, a classic example of a plant learning and remembering (Sweatt, 2016).

Mimosa can tell if it is brushed by an insect, rapidly closing its leaves, but will not close them if hit by raindrops. Plants are also aware of their neighbours, and can determine if they are the same or a different species, altering their growth patterns as necessary. Even more interestingly, wild tobacco plants, if growing next to damaged sagebrush plants, will be less damaged by an attack from herbivores than if grown beside undamaged sagebrush plants, because the sagebrush release a pheromone when attacked that the tobacco plant perceives, and then the tobacco plant produces defensive molecules, protecting itself. Plants release many such chemical signals, and these can be eavesdropped by other species, a form of wire-tapping (Karban et al., 2004).

Plants possess what is termed *dispersed intelligence*. Although they do not have a central, co-ordinating brain, they have many growing tips, called *meristems*, that act as thinking centres spread across the plant. A single winter rye plant has fourteen million roots, with fourteen million meristems, each exploring a slightly different world in terms of soil chemistry. In many ways this resembles a small version of the internet of things.

Indeed, as early as 1880, Charles Darwin wrote of the root tips of the plant as resembling the brain, co-ordinating the behaviour of the plant. He wrote that: "*It is hardly an exaggeration to say that the tip of the radicle [root] thus endowed, and having the power of directing the movements of the adjoining parts, acts like the brain of one of the lower animals*" (Darwin, 1880). The outcome is a developmental process that reflects a conversation within and beyond the plant, based on feedback.

Local solutions are essential. This is important as the above- and below-ground worlds of plants are so different and the lack of mobility as a whole requires local decision-making to take place. Rather than gathering data and sending it to a centralized brain, a plant will make its decisions where they count – at the location of the problem. The roots of the tree will generally spread around three times the diameter of the branches. But, with the fungi that form an intrinsic part of most trees, the range will be much greater, and the tree is likely to be linked up to other trees, thus becoming part of a huge network. The importance of this connectivity cannot be overstated. Here sociality is key. There is not really an individual tree, but rather an active community.

We are not dealing with individual intelligence here, but rather *social intelligence*, true community living. Imagine if our blood capillary system linked us physically to our neighbours. It would be difficult if you wanted to go somewhere where your neighbour didn't want to go, but if you are a tree, you aren't going anywhere anyway. And this is another example of how different plants are from motile animals. It is a different world with different challenges, and so the type of intelligence is completely different too.

Brainless they may be, but plants are self-organizing units, while interacting with their living and non-living surroundings. As Antony Trewavas (2017) concluded "*Intelligence…is an inevitable consequence for all organisms that consistently deal with a variable environment, both plant and animal.*"

Calvo Garzón and Keijzer (2011) have suggested that the many root tips act as eyes, sensing the local environment, with the root system displaying a swarm intelligence, wherein signals from the root tips combine to give a wide, global picture of the soil, allowing a co-ordinated response. This is a form of swarm intelligence, and takes the root-brain theory of Darwin a step further. Of course, the plant is then also part of a greater organizational whole, the ecosystem. At the end of this section we will argue that the ecosystem has its own intelligence.

The connectivity that the internet-of-things delivers, and that of AI, actually has the potential to move us more towards a dispersed intelligence model, as we become more aware of others and build common ground. Thus, plant intelligence deserves closer scrutiny, and we will integrate it with the other types of intelligence as we move towards ecosystem intelligence.

V.3. Microbial intelligence: Gene-swapping revelry in the quorum

Artificial Intelligence is usually couched as mimicking human intelligence. There is a pervading assumption that, really, only humans are actually intelligent. All the animals running around trying to eat each other, Nature raw in tooth and claw, are not truly intelligent. And also, it was the human who was made in the image of god, not gerbils or koala bears. There is no way on Earth that koala bears are made in the image of God, and they are definitely not intelligent. They live up trees, for goodness sake! Even if you widen your definition of intelligence beyond humans, it is still tightly tied to having a brain. This belief is called *cerebrocentrism* or *brain chauvinism.* If you don't have a brain, like plants and algae, then you surely can't join the intelligence club. If you are only a single cell, then definitely, no membership for you.

We like these kinds of definitions: the either-or kinds. We divide organisms into either intelligent or non-intelligent lifeforms. While we begrudgingly may admit that plants may be intelligent, something with only one cell, if classed as intelligent, would undermine not only the whole concept of intelligence, but would strip us of our sense of specialness.

Recent work on microbes suggests something different. Westerhoff et al. (2014) put it this way: "*If we were to leave terms such as human and brain out of the defining features of intelligence, all forms of life – from microbes to humans – exhibit some or all characteristics consistent with intelligence.*" Bacterial communities represent a form of social intelligence, also referred to as distributed, collective or swarm intelligence. Let's examine this claim.

All cells in the human body have the same DNA, but cells become specialized as they develop, taking on a range of roles. Some DNA is switched off, giving each type of cell a specific identity. This results in over two hundred different types of cell in our bodies. Groups of these cells form organs. This is called division of labour, where different organs take on particular tasks. However, the emergence of the multi-cellular, multi-organ being that is a human can be traced back in time. Early life consisted of single cells. Each cell had to do everything, but then they began to live together in colonies and could share out the jobs. It was like an open relationship. You aren't stuck in the one body, like cells in multicellular organisms, but you commit to doing things for each other. Many of these ancient organisms are still with us. Bacteria are an example. Bacteria can live in communities with other bacteria and fungi.

One of the most well-known of these communities is plaque. Plaque colonies grow on our teeth and need scrapped off every so often. In plaque, these different species communicate with each other, networking and making decisions.

Of course, intelligence is an emergent property, arising from organisms that are made up of lots of molecules. The individual molecules are not deemed to be intelligent. Here intelligence represents the ability to perceive the environment, to perceive change in that environment and to respond to that change in such a way as to reach a particular outcome.

What attributes of intelligence do bacteria demonstrate? Decision-making is often reduced to the "should I stay or should I go" category, but can be much more complex than that. There can be many options, all involving trade-offs and all with consequences further downstream that are not at all predictable at present. That's because we live within a system. How often have we reflected on decisions that we have made? Indeed, reflection can be viewed as an essential part of decision-making.

Do bacteria spend hours staring into space, wondering *"what if?"* You may be disappointed to read that I can't tell you that. What is clear from research is that bacteria do make decisions. They can decide whether or not to move towards a nutrient-rich spot, to move away from a toxin, and which of a range of different strategies to incorporate when short on iron (depending on the energy status of the cell). In all of these, feedback is important, very much as in human decision-making. Bacteria are also able to launch two or more different responses, at different speeds, in response to a single issue, giving greater control and adaptability, depending on the severity of the initial perturbation (He et al., 2013).

Another interesting facet is *bacterial memory*. This relies on the fact that bacteria evolve extremely quickly, partly because they can swap bits of DNA with other organisms, and partly because bacterial DNA can change very quickly. As a result, they keep coming up with novel solutions to problems. A good example is antibiotic resistance. These solutions are then kept as memories in the DNA, ready to be used when needed. This knowledge can also be taught to other bacteria by copying the pieces of DNA and then giving one copy to another bacterium through gene transfer. So, we can see that bacteria have a form of memory and of teaching and learning.

Associative learning, where we learn that something is likely to occur when something else happens, is considered an intelligent and useful form of thinking. Research has shown that bacteria can also do associative learning. Bacteria that can live in the soil but also in our guts have been shown to understand that when in the gut, oxygen levels will be low, whereas when in the soil, they will be higher. They use the temperature of their proximity to determine this. When it is around body temperature, they 'know' they are in the gut, and set in place a series of metabolic processes to cope with low oxygen (Tagkopoulos et al., 2008). At cooler temperatures (equivalent to soil temperature), the genes used to manage low oxygen conditions are switched off. Even when not in the soil or the gut, by changing the temperature in an experiment, the bacteria made the changes to respond to the oxygen levels they expected, associating temperature change with oxygen level change.

An advanced form of associative learning exists at the level of the microbial community, where *quorum sensing* operates. Here, individual cells release a signal. If there are many cells in a colony, the total amount of signal chemical will increase rapidly. These high levels of signal will then be perceived by each individual cell, and, if the levels detected are above a certain threshold, then there will be some form of response. The bacteria can work out how big the colony is. This can be seen as a form of self-awareness (another property of intelligence) at the colony level, allowing the colony to self-regulate depending on its size.

Another example of quorum sensing involves cells releasing a signal molecule if they are suffering from starvation. Once this signal reaches a threshold level, all the cells aggregate and form a reproductive structure that releases spores into the air. This allows new members of the colony to disperse to new locations, where, hopefully, food will be more available. Bacteria can also sense the signals from other species. This can be useful in a multi-species colony such as plaque.

Finally, bacteria are excellent at solving problems. For example, when exposed to antibiotics on a regular basis, such as in a hospital, bacteria quickly develop resistance. Once one bacterium solves the problem, it can pass the answer on to others of its own type and to different types, through *plasmid transfer*. In plasmid transfer, a small ring of DNA, the *plasmid*, is transferred from one bacterium to another, with the genetic solution within it. Thus, by working together and sharing solutions, the bacteria can protect themselves from attack.

I remember, as an undergraduate, working in a microbiology lab of a brilliant young professor, Hilary Kay Young. We were looking at antibiotic resistance in an estuary beside the sewage outlet of a large hospital. We took samples from the estuary and grew them in solutions containing different antibiotics. One of the estuarine bacteria had resistance to nine out of the eleven major antibiotics then available. What was amazing about this was that this particular bacterium only lived in estuaries, where the water was salty, because of the tidal effect, and would never be found in a hospital.

The bacteria in the hospital had become immune to the antibiotics, and were transported in the sewage pipe into the estuary. Here, they would die because of the salinity, but before doing so they managed to transfer their best-selling book of magic spells, entitled *Antibiotics and How to Beat Them by Working Together,* across to the estuarine bacteria, who they had never met before. Species don't exist at the bacterial level, so there is no barrier to gene-swapping, allowing them to share in the gathered knowledge, and to rapidly adapt to any problems they face. It's why the bacteria have been around for nearly four billion years.

Shared intelligence is better than secretive intelligence. At least that's what they say. This reminds us of the issues surrounding *big data* and *data protection*. If everyone can access everything, then there is no inequality and complete transparency. A problem solved and shared is a bigger problem for your opponent, especially if you were a human canoeist, paddling near that sewage outlet. But when you are all in the same boat, you don't worry about it. For a young student though, it was very exciting.

At the heart of this bacterial intelligence lies feedback and communication. The bacteria are critically aware of the narrative around them, and contribute to this conversation, awaiting cues that inform and trigger their decision-making. Whole stories are transferred between them, much like the story-telling traditions of ancient civilizations. Hula dancers in Hawaii are not dancers at all, but recite the stories of their ancestors to music and dance.

I remember walking in the dusty, hot Djemaa el-Fnaa Square in Marrakech, Morocco. Avoiding the incessant approaches of the market traders, I noticed a man with an ever-growing crowd around him. They were transfixed as he swayed and spoke in a hypnotic way. I couldn't move away. I didn't understand anything he said, but knew that this was a powerful moment. I later learnt that he was a story-teller who had come down from the Atlas Mountains to share his tales, educating, thrilling and inspiring his audience. That was twenty years ago. Today, hardly a story-teller can be found in Marrakech, a sad reflection of a lost art, and so much more loss.

The transfer of stories, in the form of plasmids, is alive and well in the bacterial world, providing inspiration, education and new solution spaces, as our one-celled friends continue to thrive. Story-telling leads to problem-solving, much like it did in our ancient ancestors. In colonies made up of different types of organisms, such as plaque, communication between these species takes place, and each type of organism will take on particular roles for the good of the colony. It should be remembered that a colony of plaque on your teeth is an extremely complicated society, with maybe one hundred different types of bacteria, each with ten billion individuals, totalling over one hundred times the planet's human population and involving a huge network of communication (Jacob et al., 2004), all living in your mouth.

Such inter-specific communication is something humans encounter only in Dr Doolittle in the series of books by Hugh Lofting, or *The Horse Whisperer* by Nicholas Evans. Communication in Nature is often in the form of chemical exchange, wherein messages, with very specific meaning, can be transduced. These chemical signals can last for days or weeks, acting as signposts and maps. Of course, chemical signalling is extremely important in most organisms, but we are largely unaware of its intricacies, due to the poor development of our sense of smell. We, and other higher primates, have 'dry' noses, meaning we miss many of the chemical signals that other animals can smell, with their 'wet' noses, such as cats and dogs.

While today, humans have access to so much information through books and multi-media, we have become numb to the natural environment and the conversations to be had there. The rest of Nature is heavily directed by these conversations. The competitive Darwinian approach we have taken is not that of the bacterial world, of which Darwin knew nothing. And the social intelligence of micro-organisms is extremely advanced in many ways. It just works differently to how we understand intelligence. Society, from a microbial point of view, is an essential element of survival, not merely an optional extra. It used to be like this for humans too, and, in many parts of the world, it still is.

A fascinating recent discovery is that the human intestine acts as a second brain, and is thought to influence behavioural traits. The gut-brain axis links the intestine to the cognitive and emotional centres of the brain (Carabotti et al., 2015), where metabolites synthesized by the bacteria living in the gut impact on brain function. This has become a topic of increasing research. It is bi-directional. If the microbial community in the human gut is somehow destabilized, through for example, a course of antibiotics, or overuse of detergents in a baby's surroundings, the consequences for our mental health can be significant (Rogers et al., 2016). Research has shown that such destabilization contributes to autism spectrum disorder, schizophrenia, depressive disorders, Parkinson's disease and anxiety (Sudo et al., 2004; Dinan et al., 2013; Mayer et al., 2014).

The average human gut has around one hundred million million bacteria (Tannock, 1995), which is ten times more than there are human cells in the body. By cell count, we are only ten percent human and ninety percent bacterial (not counting the mitochondria (the cell batteries) in each human cell, which are the remnants of ancient bacteria themselves (Dyall et al., 2004)). This gut microflora is crucial to central nervous system development in new-born babies and also fine tunes their immune systems.

So, we see that the bacteria living in our guts have an impact upon our mental health, decision-making and functionality through chemical signalling and community structure. Thus, microbial activity is influencing our intellectual functionality. The gut is a second brain, and much of its thinking is done by the bacteria living there.

It is now time to examine a higher intelligence, acting as the conductor of the orchestra. The different forms of intelligence that we have encountered in this section all play significant roles, but the Earth system itself is a co-ordinated entity, and possesses ecosystem intelligence. This book suggests that ecosystem intelligence, rather than human intelligence, makes the best model on which to base artificial intelligence, basically because it has proved itself over three billion years, running the most complex and multifaceted show on Earth, the living planet.

V.4. Ecosystem intelligence: Systems thinking in the cathedral of thought

With artificial intelligence promising to play a central role in human activity in the coming years, its relationship with sustainability comes to the fore. Humans are facing unprecedented challenges, mostly of their own making, that are existential in scale. As we have seen, the Enlightenment forged a philosophy of human-centric focus, wherein the betterment of humankind was to be our core goal. Progress, economic growth and development encapsulate this brave new world, where the planet would serve us and form the source and sink of our activities, no longer threatening us.

The Malthusian horsemen of the apocalypse were to be put out to pasture. Humans were intelligent, whereas Nature was raw in tooth and claw, chaotic, murderous and not very bright. Nature relies on us for conservation, and is really only useful as a resource and somewhere for us to take the kids on a day out. There is nothing in Nature for us to learn from. As Condorcet said, Nature won't limit the progress of humans.

Yet Nature has managed just fine for over three billion years without our intervention. Mass extinctions came and went, as did snowball Earth, when most of the planet froze over. Giant asteroids as large as Mount Everest impacted the planet. Huge volcanic eruptions covered areas the size of Siberia and India with vast sheets of lava. Somehow, big, stupid Nature was able to pick itself up off the canvas without our help each time.

Something as sophisticated and complex as the living planet takes some re-organizing, especially when recovering from devastation. However, Nature has recovered all on its own, producing forests that need no management, savannahs and flowering prairies that need no fire policy (because fire is a central part of their ecology) and oceans that needed no marine conservation plan. Nature's garden represents a beautifully dynamic equilibrium, everything in its place and yet constantly changing.

And so, when it comes to managing our own survival and the sustainability of the planet, there already is an intelligence capable of doing the job. Why re-invent the wheel?

We don't need human intelligence, but *ecosystem intelligence*. Ecosystem intelligence underpins the structure and functioning of our planet. It operates seamlessly at every level of organization throughout the planetary system. It is making decisions all of the time, monitoring, listening and responding. It has a set of rules and insight better than 20/20 vision. It reaches into every aspect of the biosphere, from the sub-atomic to the circulatory patterns in the atmosphere and the oceans.

Within the Earth system are a wide range of intelligences, as we have seen, from bacterial intelligence, with their plasmid transfer, quorum sensing, associative learning and inter-specific communication, to plant intelligence with its dispersed intelligence and social connectivity with their own and other species through fungal links. We've encountered swarm intelligence, Machiavellian intelligence, and complex tradition in animals. Oh, and then there is human intelligence. And all of these different ways of thinking contribute to the Earth system in particular ways. But infusing all of this is ecosystem intelligence.

Parallel living is not possible, where we do our thing and the rest of the planet does its thing. Rather, we need to re-integrate. By this I mean that we need to work within the Earth system, participating and playing to the rules. This doesn't mean we need to start adopting the intelligence of a plant or bacterium, even if this was possible. For the Earth system is a diverse church and all of the congregation are welcome to bring their thoughts to the table.

Crucially, however, we need to be in resonance with the rest of the system, rather than separating from it and enforcing our thoughts upon everything else. Otherwise, we will lose our place on the field of play, red-carded, having accumulated too many transgressions. The changing room of extinction awaits.

It's happened before, many millions of times. Species come and species go. It will continue to happen. We've only been here for maybe three hundred thousand years as anatomically modern *Homo sapiens*. Cow sharks have been around for one hundred and seventy-five million years. Horsetail ferns first appeared around three hundred and fifty million years ago, and still grow along the roadsides in our village. Horseshoe crabs can trace their history back four hundred and twenty-five million years. Velvet worms may date back five hundred million years.

We are newbies, and the most vulnerable of all. We are warm-blooded, so require a huge amount of energy to stay alive. We are more vulnerable to just about everything than any other form of life. This vulnerability stems from two things. Firstly, we reproduce via sexual reproduction, meaning we can't swap genes with other species like bacteria are able to do. We also reproduce slowly, and, usually, with the one partner, thus greatly limiting the genetic variation that we can generate in the next generation.

Secondly, we require complex niches in which to live. Things have to be just about perfect for us. We get food poisoning easily, and die. If our body temperature drops by two degrees centigrade, we fall into hypothermia, any lower and we will die. If our temperature rises seven degrees above normal body temperature, we die of hyperthermia. We need to be adequately hydrated. If we lose between fifteen and twenty five percent of our body's water, we will die. For a comparison, many algae, such as nori (*Porphyra*) can lose ninety eight percent of their water and survive (Skene, 2004). If our blood salt levels increase by only sixty milligrams per litre of blood, we will die from salt poisoning. Infection from just a scratch from a rose bush can kill us.

We are much less *Homo sapiens* than *Homo vulnerabilis*. Bacteria can survive much more easily as they need less energy and have strategies to cope with all eventualities. In 2007, bacteria found in permafrost in Canada were estimated to be five hundred thousand years old (Johnson et al., 2007). *Geogemma barossii*, a bacterial species that lives on hydrothermal vents, where life on Earth may have begun, can live and reproduce at one hundred and twenty degrees centigrade or two hundred and fifty degrees Fahrenheit (Frappier and Najmanovich, 2015).

So, what is ecosystem intelligence? To understand this, we need to understand what an ecosystem is. The etymology of "ecosystem" is extremely informative. The term *'eco-'* comes from the Ancient Greek word οἶκος (*oikos*), meaning *'household'*. The word *'system'* comes from the Ancient Greek σύστημα (*systēma*) meaning *'a composition'*. Thus, ecosystem literally means the composition of the household.

It's a bit like an orchestra, whose members each occupy different rooms in the same house, where each individual may only directly hear the few other players in surrounding rooms. Yet together the players produce a mesmeric musical performance. In much the same way, an ecosystem, made of many millions of bacteria, fungi, algae, plants and animals, each of which interacts directly with only a few others, produces a complex system in equilibrium, functioning as an intact super-organism. To understand how this works, we need to understand the properties of systems more generally.

A system is a network of interconnected, mutually dependent parts that together form a unified whole. The word has entered into common parlance. According to the *espressoenglish* website (2019), of the thousand most commonly used words in the English language, the word *system* comes in at 192nd, much higher than information (316th), education (378th), economy (647th), nature (698th) or environment (859th). Systems have four key characteristics when functioning properly, and it is essential that we understand these if we are to find a way out of the mess that we are in. These characteristics allow the system to function, to recover and to maintain itself through time. They also set out the rules of the club that all members must abide by. Let's have a look at each of these in turn.

V.5. Systems are non-linear

Because of the many players within a system, the system as a whole does not travel in a direction from a departure station to a destination station. It is not like a train. Rather the journey is dynamic, and there is no railway track. To understand this, we can look at what has happened in physics over the last four hundred years or so.

Newtonian physics, stemming from the work of Isaac Newton (1643-1727), relies on the foundation of cause and effect. You could predict what would happen to a projectile if you knew the angle it was thrown at and the initial speed of the object (i.e. how hard it was thrown). Every event is caused by a previous event and so on, forming a chain of events. Cause leads to effect through time. The apple will always fall from the tree. The material reality and the laws controlling that reality are predictable. A plus B always equals C. This approach is called *scientific determinism* and is a cornerstone of empiricism.

However, physics underwent a velvet revolution in the early twentieth century. *The New Physics* introduced the *Uncertainty Principle*, first defined by Werner Heisenberg, which states that the speed and position of the tiny particles that make up our atoms cannot both be known at the same time. Also, any attempt made to measure these two properties will alter the object being observed (the observer effect).

Why is this so important? Simply put, observation disturbs reality. Since we can't be sure of the present reality, we can be less sure of the future reality. Thus, the sacred bond between observation and the material world is broken. Einstein, a major protagonist of the New Physics, stated: "*Even space and time are forms of intuition, which can no more be divorced from consciousness than can our concept of colour or shape or size. Space has no objective reality except as an order or arrangement of objects we perceive in it, and time has no independent existence apart from the order of events by which we measure it*" (in Barnett, 1949).

Fundamental to all of this is the idea that perception is dependent on context, and that perception and context interact. The Cosmos is non-linear, operating as a giant interactive system. A butterfly's wings, flapping in a distant forest, may appear to go unheard, but the flapping will ultimately contribute in multiple ways, to everything else that happens. Just because we can't determine cause and effect doesn't mean that change doesn't happen. This is the essence of non-linearity. And it makes life a lot more interesting too. Reality is a wicked problem and not a brick-building exercise.

V.6. Systems are emergent

In any complex system, interactions of the parts give rise to properties that only belong to the whole. In other words, one plus one does not always equal two. This is called *emergence*. Emergent entities arise out of more fundamental entities but cannot be reduced to these entities. One of the earliest philosophers to work on emergence, John Mill, wrote that "*To whatever degree we might imagine our knowledge of the properties of the several ingredients of a living body to be extended and perfected, it is certain that no mere summing up of the separate actions of those elements will ever amount to the action of the living body itself*" (Mill, 1843, Bk.III, Ch.6, line 1).

There are lots of examples of emergence. Culture is one. The punk movement, the hippy movement, the Aztec civilization and the Incans are all examples of emergent cultures. It is impossible to predict what the next culture will look like, when it will emerge or from where it will come. The whole is greater than the parts and comes from components that don't, in themselves, represent the culture. Culture has many facets that all interact, such as fashion, music, philosophy, landscape, art, architecture, poetry, dance, nutrition and, often, religion.

Another example of emergence is our dream world. Drifting off for a night's sleep, maybe after a few glasses of Vino Nobile di Montepulciano and some Pienza cheese, we enter the altered universe of our dreams. While they can be linked to events in our past, dreams still oftentimes explore an unexpected and surreal landscape. Yet our brains are made up of fairly basic neurone cells connected together by fairly basic chemicals, none of which on their own could produce such a panoply of imagined experiences. Indeed, creativity in the waking world is also emergent, particularly in the genres of fantasy and surreal art.

Emergence embodies unpredictability, an indeterminate destination and a dynamism that creates significant challenges to control freaks such as humans. To exist within a system, we have to relinquish control and embrace the unknown. The emphasis is on process and function, not form and outcome. If the processes are working well, then the system will reveal its path as we travel it. But the path only appears as our feet touch the ground.

The road ahead is invisible to us. It requires some form of faith to keep walking forward. However, this is not the walk of Schopenhauer's crippled victim, who can see the light, but who is carried by a blind giant representing his will. Rather, the journey unfolds as an emergent outplaying of the interactions of all those within the system, connected and alive to the context that is the Earth system.

And so, a system does not have a sense of progress towards a particular structural destination, but rather it exists within a dynamic equilibrium. It does its thing and then outcomes emerge. Goals, forms, objectives, timelines and targets are so central to all that the modern human defines as measures of progress, especially in the West and North. But in a truly functional system, these concepts are not important. Emergence is the opposite of reductionism, wherein something is built from simple building blocks.

In other words, you cannot reduce reality to these blocks. In modern biology, genes are the building blocks, and everything else comes from these basic units of life. This concept means you can alter things by adding or subtracting genes, in what is called genetic engineering. You can add a gene for disease resistance, and the crop will become immune to that disease. You can add a gene for herbicide resistance, and then spray the crop with herbicide, killing all the weeds, but not harming the crop.

Unfortunately, Nature doesn't play reductionism very well. Disease organisms overcome the resistance programmed into the crop. Weeds become resistant to the herbicide. And the more you use these approaches, the greater the resistance becomes. It's a lot like antibiotic resistance. That's because Nature is much more complex than a simple set of building blocks. Things impact on other things because Nature is relational. Everything effects everything else. And there are a whole host of different types of intelligence, as we have seen, all very capable of solving problems quickly and differently.

There are so many interactions that we cannot predict exactly what will happen. This is the essence of emergence.

Reductionists don't like to let go though. A rather amusing example is in a paper by Hiroaki Kitano, in the Journal, *Science*. With the promising title, *Systems biology: a brief overview,* Kitano exposes his reductionist roots when he comments *"There is now a golden opportunity for system-level analysis to be grounded in molecular-level understanding"* (Kitano, 2002). He may wish this was the case, but his *"golden opportunity"* will turn to clay, in a reversal of alchemy, once he fully embraces systems theory.

The Earth system is not grounded in molecular level foundations, but in emergence, wherein the complexities of interactions determine what the outcome will be. Molecular biology is only one level of organization in the greater scheme of things, but try explaining forest succession through molecular biology. It is exactly this reductionist attitude, so clearly represented in what Kitano says, that prevents us recognizing what really needs to be done.

V.7. Systems are sub-optimal

If emergence is challenging to us, the third property of systems, *sub-optimality*, is perhaps even more so. All systems are sub-optimal at the level of the individual components. In other words, to quote the Rolling Stones, *"You can't always get what you want"*. The reason for this is because there have to be trade-offs. It is impossible to please everybody all of the time, as any human resources manager will tell you. Each component in a system has an ideal world where everything is perfect for them. However, the likelihood is that each set of individual interests will clash with others.

Take, for example, the design for a car. The lighter it is, the less power is needed to accelerate. However, strength is needed to prevent significant damage in the case of a crash. Also, all the extra luxuries demand power to run them and they increase the weight further. A wedge-shaped car is more thermodynamic, but a family car needs plenty of room. All of these properties compromise each other and they can't all be optimized.

Nature is sub-optimal. Foxes don't eat all the rabbits. Rabbits don't manage to evade all the foxes. DNA doesn't perfectly correct itself from damage, allowing some mutations to occur. This provides genetic variation among populations. This variation means that if, for example, a virus attacked the human race, at least some of us would be resistant. If we were all identical, with no mutations, then we would not have this variation, and would not have variable resistance. However, we do correct most mutations, avoiding our DNA from becoming so altered that it cannot function.

Squirrels don't remember where they hid all of their nut stashes, allowing some of the nuts to germinate and grow into trees, thus providing nuts for future generations. Also, bears are able to find some of the forgotten stashes, helping them to survive winter. However. the squirrels remember where enough nuts are located so that the squirrels themselves survive. Meanwhile many tree species reduce nut production for a number of years, controlling squirrel populations, and then have a mast year, where they produce vast amounts of nuts, ensuring there are more nuts than the squirrels can hide. This is a good back-up plan.

Imagine if squirrels had a nut app. Each year they would carefully log where every stash of nuts was buried, allowing them to optimally harvest the nuts. There would be no new trees and eventually the forest would die away. There would then be no nuts, meaning the squirrels' future generations would be compromised. With all the nuts now accessible, the squirrel population would start to increase. While this is good for nut app sales, it would put more pressure on the nut harvest. Squirrels also eat bird eggs in the spring for extra protein, and so with increasing squirrel populations, more bird eggs would be eaten.

The squirrels might even invest in further technology, an egg app, to help find all the bird nests. This would lead to the extinction of the bird populations. This, in turn, would lead to insect populations growing out of control, since there are no longer any birds to eat the insects. Excessive numbers of insect larvae would mean excessive damage to foliage, massively impacting trees and shrubs, and depressing flower and fruit production, as there would be no longer sufficient sugar produced through photosynthesis to support the production of nectar and fruit. This fruit collapse would impact frugivores (fruit eaters) and bees, who rely on the nectar. With the bees in crisis, pollination more generally would collapse.

As forests decline due to the nut app improving the efficiency of the squirrels' harvesting efforts, and the egg app leading to the collapse of bird species and an exponential increase in leaf damage from insect populations now spiralling out of control, soil erosion would increase and the water holding capacity and evaporation of water through the trees of the ecosystem would decline, meaning that rainwater would more quickly flow downstream causing flooding.

With pollination declining, food supply would decline considerably. Thus, key ecosystem functioning would collapse. With bears starving in the winter, their roles in spreading fruit seeds through their droppings, and in controlling populations of their prey, such as deer, elk and fish, would become hugely threatened. Bears are generally keystone species, meaning they play highly significant roles in the shaping and maintenance of ecosystems. Thus, if you mess with the bears, you mess with the whole system.

As can be seen, by optimising for the squirrel, using advanced technologies such as nut and egg apps, the entire ecosystem would likely collapse. Nature relies on sub-optimality in order to function properly. Any functioning system is the same. And this is the fundamental problem with humans.

For humans have focused on optimising natural processes for their own gains. The optimization of agricultural productivity is a classic point. The industrialization of growing plants and animals has led to many of the significant problems facing us today. A combination of forest-clearing for agricultural purposes, having excessive browsing and grazing animals on the land and poor crop rotation means soil erosion has soared. Excessive irrigation, particularly in semi-arid regions of the world, where farmers try to grow crops in climates not conducive to this, has led to salinization of the soil.

In order to further optimize agricultural productivity, we have added huge amounts of fertilizers, such as nitrogen, potassium, iron and phosphorus, to the soils. Much of this fertilizer then is washed into the rivers, lakes and coastal waters, leading to toxic cyanobacterial blooms, fish death and ecosystem collapse.

Finally, agriculture contributes more to climate destabilization than any other industry. Whether it is methane released from excessively inflated cattle populations, or carbon dioxide from the hugely energy-expensive Haber-Bosch process of nitrogen fertilizer production, greenhouse gases are produced due to our optimization of food production.

The combine harvester is another example of efficiency, hoovering up every last grain from the vast fields of cereal crops. In agrarian societies of old, harvesting by scythe and hand bailing meant that lots of grain was left behind, the angels' share, akin to the whiskey industry, that would feed the birds and small mammals. These in turn would eat insects and control pest species. Now, many of these bird and mammal species, such as the humble sparrow are threatened to extinction.

Mao Zedong, the founder of the People's Republic of China, learnt to his cost how important the sparrow was. Mao had heard that sparrows ate a lot of grain, and so in 1958, Mao ordered that all sparrows should be killed as part of his *Great Leap Forward* programme of change. The *Great Sparrow Campaign* was launched that year, and hundreds of millions of sparrows were duly dispatched by a population eager to please. The campaign was deemed a grand success.

It wasn't until two years later that problems became apparent. You see, sparrows also eat the larvae of the locust. The absence of sparrows meant a locust population explosion. The locusts swarmed, eating most of the crops. It has been estimated that between fifteen and seventy-eight million people may have starved to death in the great famine that resulted. Today, the threat posed to these seed-eating birds comes not from some government-sponsored killing spree, but from improved efficiency, or optimization, in harvesting, wherein there is not sufficient grain left over to sustain these populations.

Although the idea of a squirrel using a nut app or an egg app is amusing, the point is serious. By optimizing for humans, we are undermining an essential rule for system-living: sub-optimality. Optimizing for our own progress is leading to the collapse of the Earth system upon which we rely. From the onset of the agricultural revolution some twelve thousand years ago, we have continued on our path of optimization. And this is why our problems are escalating.

A final point on sub-optimality relates to the decision-making on trade-offs. Amongst humans, things like trade negotiations can go on for years. The Banana Trade Agreement, signed between the EU and Latin American nations in 2009 had taken eighteen years to complete. A free trade agreement between the USA and Panama took eight and a half years from launch to implementation. This is because humans like to assert, or at least be seen to assert, control during disputes over ownership in any negotiation.

When the Sustainable Development Goals (SDGs) were launched by the UN in 2015, the interconnectedness of the goals was emphasised. But with one hundred and sixty-nine targets and three hundred and four indicators, who is going to elucidate the required trade-off balance between these multiple aims? For example, how do we balance Goal Eight (decent work and economic growth) with Goal Thirteen (climate action), or Goals Fourteen and Fifteen (life below water and on land) with Goal Two (zero hunger)?

And who makes this decision? In the natural system, there isn't a small committee who decides. Here, trade-offs are not a matter of who should make the decision. Rather, it is all about what effects will the trade-off have. The balance of power lies with the system, not with any particular individual. Thus, the outcomes are emergent and sub-optimal, with interconnectivity and feedback playing key roles. The parties involved in discussing the trade-offs don't just meet every three months in Nature. They cohabit the same room, and so are able to find the balance much more quickly. Perhaps negotiators in human trade deals should do the same.

So how sub-optimal should we be? It's an interesting question. Returning to our squirrels, there is a balance between forgetting where any of the nut stashes are hidden, and recalling where every last nut is buried. So how do we know how sub-optimal to be? It will depend on a number of things, and, more fundamentally, must be a dynamic response. By this I mean that there is not a fixed level of sub-optimality. Depending on the context, we may need to be more or less sub-optimal, and it is a forever-changing conversation. We are not building a static structure, but rather, working within an emergent reality that depends upon real-time responsiveness. It's a bit like the concept of *carrying capacity*.

In 1588, the Italian polymath, Giovanni Botero, writing in his epic tome, *On the Causes of the Greatness of Cities*, demonstrated that a city existed as a balance between reproduction and nutrition. Both factors were variable and so the size of the city could be expected to change over time. It was the first expression of the idea of carrying capacity, the number of a particular organism that can be supported, or carried, by a given habitat.

It would be a huge mistake to think this number is constant. As a habitat changes, so does the carrying capacity. Thus, if you keep a herd of cattle of the same size in a particular field, as the field degrades, due to vegetation being eaten, soil being disturbed by hooves and erosion of this exposed soil, the carrying capacity will diminish, and the field will become increasingly degraded due to the cattle population exceeding the capacity of the habitat to support it. The tragedy of the commons doesn't necessarily need an increase in the herd size. A degradation of the field can be equally tragic.

The term, carrying capacity, actually was first used not in ecology but in merchant shipping. Before 1845, tax was levied on ship cargo based on a calculation of the volume of the ship depending upon the length, depth and breadth of the boat under consideration. However, with the introduction of steam-powered ships, the owners of these new forms of transport complained that some of their volume was taken up by engines, whereas sailing boats did not have such wasted volume. Therefore, according to the steamer captains, they were being taxed for more cargo than they actually contained, putting them at an economic disadvantage. On 13th January, 1845, the concept of carrying capacity was introduced wherein the volume available for carrying cargo for each ship was measured individually rather than basing it on overall dimensions.

Modern agriculture in many ways is all about carrying capacities. We seek to reduce the carrying capacity of the habitat for competitors (such as weeds) and pests (such as insects), herbivores (such as rabbits), and carnivores (such as wolves, if we have livestock), while maximizing carrying capacity for crops and livestock.

However, all of this habitat manipulation is actually related to another carrying capacity: our own. Because ultimately, we are seeking to optimize our own success. But at what an expense. We use fertilizers, pesticides and insecticides, at huge cost to our environment. Remember Mao Zedong and his Great Sparrow Campaign? Intensive farming leads to soil erosion and salinization, meaning we have to add even more technology just to stand still in our fight for food, a fight we picked in the first place.

So how inefficient do we need to become? Interesting, there is a theory in ecology to help us: the *intermediate disturbance hypothesis*. Disturbance is a bit like sub-optimality. You need some of it for a diverse ecosystem, but not too much. Too much disturbance and everything dies, but a middling amount, a bit like Goldilocks, and there is enough change to allow more species to co-exist. No disturbance at all, and a few species will dominate. Disturbance ensures a dynamic situation. In order to work out how much disturbance you need, you should increase it until diversity peaks. After this point, there is too much chaos, and before it there isn't enough.

The same applies to sub-optimality. When any given system is working at its best, then each individual component is operating at the right level of sub-optimality. I like to imagine a system as a giant mixing desk you find in a recording studio. Each individual instrument can be recorded on a track, and then all of the separate tracks can be mixed, by turning knobs on the desk, so that together the balance is perfect: just the right amount of bass, drums, guitar and vocals. Of course, if the guitarist was in charge of the mixing, you probably wouldn't hear another instrument. He would seek to maximize the volume of the guitar. Mixing takes a system approach, targeting the best overall sound.

Of course, like carrying capacity, there will not be a static amount of sub-optimality in any given activity. Rather, since the entire system is dynamic, then we would expect the appropriate level of sub-optimality to vary. Optimization for ourselves lies at the heart of many of the environmental problems facing us today. This key characteristic of systems is clearly central to any hope of sustainability.

So, when each component is contributing to the whole in the most appropriate way, the whole system takes on a quality that has maximum resilience, resistance and functionality. We need to examine in detail the impact of our actions and determine when they are appropriately sub-optimal. The only way to do this is to monitor the entire system.

To monitor the Earth system, we need to have some form of measurement. But how are we meant to monitor something so huge? It just so happens that, ironically, while we are at a point in our history where we are almost completely deaf to what our environment is telling us, we also live in a time where humans have greater capabilities of listening than ever before. The internet of things is very good at listening. With seven billion devices scattered across the globe and above it in the form of satellites, and with AI systems capable of analyzing this flow of live data, we've never had it so good. Technology is all around us, and above us. Artificial intelligence can be the powerful mediator of this process of re-unification and sustainability, but if it is to play this role, it must incorporate ecosystem intelligence, not human intelligence as its model. And this brings us to the final key characteristic of systems: feedback.

V.8. Systems rely on real-time feedback

The final characteristic, which lies at the heart of any successful system, is real-time feedback. All of the components of a healthy functioning system are constantly communicating with each other across different levels of organization. This communication is in two forms: direct and indirect feedback. *Direct feedback* occurs between two closely-linked components.

For example, predator and prey population numbers hugely affect each other. If prey numbers increase, then predator numbers increase, while if predator numbers increase, prey numbers start to decrease. Decreasing prey numbers then lead to decreasing predator numbers. We end up with two waves chasing each other. The link is simple. Too few prey items, and the predator starves and doesn't produce many offspring, while too few predators and the prey population can increase as there are less predators to eat them. As Woody Allen famously wrote, in the movie script for *Love and Death, "To me nature is…spiders and bugs, and big fish eating little fish, and plants eating plants, and animals eating… It's like an enormous restaurant, that's the way I see it."*

Habitat size and quality greatly impact on population size, as does water availability, maximum and minimum temperature and altitude. With temperatures rising across the planet, species are moving. Some species grow higher up mountains than before, because it is now warmer at higher levels. However, these species displace other species that lived there before, and eventually this shift in species forces some to the top of the mountain, from where they have nowhere left to go. In the Swiss Alps, research has shown that plants are shifting upwards sixty feet (twenty metres) per decade.

Meanwhile in Antarctica, krill populations have shifted south by two hundred and seventy miles (four hundred and forty kilometres) over the last ninety years, their distribution having shrunk, and there are less new krill being produced. This is already having a significant impact upon whales, seals and penguins that rely on krill as their major food source. Furthermore, the further south the krill move, the less phytoplankton are available in autumn and winter because of the decreasing daylength. Thus, as increasing sea temperatures drive the krill towards the South Pole in search of cooler water, it also drives them to starvation in winter.

This is interesting. You might think that as the sea warms, the krill can just shift to cooler water nearer the South Pole. However, although this solves the temperature problem, you can't change the length of the seasons. Darkness still occurs in winter for twenty-four hours each day, and so the little algae that the krill need for food can't grow without light.

Indirect effects occur when changes in apparently unconnected processes impact on particular components. In 1926, wolves became extinct in Yellowstone National Park. Following a re-introduction programme in 1995, elk populations began to decline. The elks avoided open regions such as the valleys where the rivers flowed, for fear of predation. Aspen and willows began to grow again along the riversides, now that the elk no longer grazed there. These trees and plants stabilized the river banks.

Stabilised banks meant the rivers meandered less, and formed large pools all of which encouraged life in the water. They also attracted back many bird species, who nested in the newly grown trees. Because trees are key water conduits, moving water from the soil to the atmosphere through their leaves, and because of their extensive root systems, soil erosion declined. More shrubs meant more fruit, allowing bears and other frugivores to thrive. The forested areas also provided greater shelter (Winnie Jr and Creel, 2017). Of course, the elk took a hit (from the wolves and bears), but the system became more diverse, resilient and resistant.

Another example was the re-introduction of sea otters on the west coast of Vancouver Island in 1969, forty years after the last sightings (Markel and Shurin, 2015). The sea otters reduced the sea urchin numbers massively. This allowed the kelp forests to grow much larger. The kelp provided food and structural habitat, leading to a rapid increase in fish and invertebrate species. The kelp also moderated current flow. This provided shelter for many species and allowed mussel and barnacle populations to thrive, also sheltering shores by reducing wave energy. Another interesting impact was that rock fish moved up the trophic ladder, by eating mostly fish rather than invertebrates. This change in trophic structure is another example of an indirect effect.

Indirect effects are not always good. Ironically, one example of this was a classic piece of reductionist thinking, when cane toads from South America were introduced into Queensland (Australia) to control pests of sugar cane, an important crop of the region. Cane toads have poisonous skins, and many vertebrate predator species in Australia have seen sharp declines in their populations as a result of eating the toads, including fresh water crocodiles, snakes and monitor lizards. As a result, smaller lizards, normally predated by these species, are now increasing in numbers, greatly effecting their prey species (Feit et al., 2018).

Temperature, nutrient levels, UV, changing rainfall patterns, water availability, habitat patch size, food chain changes and disturbance all have impacts, directly or indirectly, on ecosystems. Ecosystems are always listening and a properly functioning system responds to the multitude of messages rapidly. This may involve changes in population size, moving location or altering physiology. However, the response is often limited by the diversity within the ecosystem. Ecosystems with lots of species can best respond to changing messages because they have a greater variety of ways to do this.

Furthermore, individual populations with lots of variation within them can also help, again because they have a greater range of responses available. It is like having a quiz team with ten members instead of just one. It is more likely that the team of ten will be able to answer a random question than the team of one because, combined, they will have a greater number of skills and expertise. Diversity provides resilience and resistance, presenting a larger solution space within which to solve problems.

Feedback cycles, where detection of issues and responses to these issues occurs in real-time, are central to a functioning Earth system. They allow the system to respond to change, and lead to a dynamic equilibrium. It is the process of feedback that underpins the other characteristics of systems (non-linear response, emergence and sub-optimality). Constant data gathering and flow means that a truly vibrant system state exists.

And so, we find ourselves faced with the natural world, that has been refining and developing its system skills for over three billion years. We emerged from that world and for ninety nine percent of our time on Earth, our sub-species has lived within this system, as hunter gatherers and omnivores.

We have contributed to and been directed by the system. But latterly, we have sought to separate ourselves and optimize things for the human race, breaching the system code of conduct. We have damaged the system, but more fundamentally, jeopardized our own hopes for a sustainable future. The natural world will carry on without us, as it did for aeons before. It can overcome the destruction we have unleashed upon the planet, as it has through ages past.

But if we are to preserve our place in the scheme of things, there is only one way, and that is to re-integrate into the system. We need to escape from bubble world, the isolationist and human-centric place within which we find ourselves (Skene, 2011). And if AI truly represents our future, then AI must be rooted not in the Enlightenment intellect of humanity, but in ecological intelligence, system-based and founded in the four key characteristics of non-linearity, emergence, sub-optimality and real-time feedback.

This book sets out this fundamental rebooting as essential for our survival.

It is now time to examine the consequences of our journey on Earth, and the damage done. We must face this in order to understand the urgency and necessity of change.

SECTION VI Highway to Hell: The Existentialist Threat Facing Humankind

"The great increase of our powers is itself maybe the most immediate cause of our loss of vision. It must be a sort of natural law that any increase in man's strength must involve a lengthening of his shadow; as we grow in power we are pursued by an ever-growing darkness…Power has darkened us. The greater the power grows, the harder it is for us to see beyond it, or to see the alternatives to it. It exercises as compelling an influence on us, who possess and use it, as it does on those we use it upon and against."

–Wendell Berry, *The Long-Legged House* (Berry, 2012 [originally 1967])

Welcome to Section VI. Having explored the field of AI and examined alternative models to human intelligence, we now turn to the greatest challenge facing the human race at present, our very survival. This is important because it sets out the context within which AI can find its transformative role, one for which it was not created, but for which it is ideally suited: saving the human race from extinction.

We open with an overview of our history on Earth. It is an extraordinary story, where, in twelve thousand years, we have transformed our planet into a source and sink, and re-aligned it to satiate our needs and desires. Urban and agricultural ecosystems now dominate the planet. We examine the consequences of our industrial legacy, focusing on the environment and society.

We identify our five greatest impacts on the planet, and discover that they are interconnected. These five impacts have devastating effects, not only on the Earth system on which we depend, but on our societies. To understand how we have arrived at such a self-destructive point in our history, we go back to the work of one of the greatest Enlightenment thinkers, Adam Smith. We discover a man whose work was so devastatingly mis-interpreted and then abused as the basis of our current economic framework, delivering such calamity upon us all.

We then encounter Simon Kuznets and his seriously flawed graph that has been adapted as the bulwark of current sustainable development thinking, while, in reality, it drags us closer to the edge.

VI.1. A brief history of our path towards destruction

Our journey has been quite extraordinary. Within twelve thousand years we have gone from being hunter gatherers, most likely still living alongside the remnant hybrid forms of more primitive members of our genus, such as the Neanderthals, to a planet heavily dependent on computer-based technology, with hundreds of satellites orbiting the globe and nuclear power plants generating untold energy from splitting atoms of uranium.

Over these twelve thousand years we have built cities, developed agriculture, mined fossil fuels, and changed the biosphere, atmosphere, hydrosphere and geosphere beyond recognition. Harari (2017) points out that today, more people commit suicide than die at the hands of war, more people die of over-eating than from famine and more people die of old age than infectious disease. The changes have been so significant that they constitute a new geological epoch, the Anthropocene.

Across this period, our population has risen from around two million people to seven and a half billion. We have taken over the planet to produce the food and luxury that we feel entitled to, using Nature as a source and a sink. But the system is starting to creek. To understand what is needed to alter the path of destruction that we find ourselves on, we need to understand our changing relationship with the three major arenas of human activity: environment, society and economics.

The environmental arena consists of the water (hydrosphere), land (geosphere) and air (atmosphere), and everything that lives in them (the biosphere). Within the environment, water, nutrients and carbon cycle around and between these four spheres. Of extreme importance is the rate of flow of this cycling. Diversity, resilience and resistance are all interactive environmental characters, reflecting the health of the system. Energy enters in the form of sunlight and is dissipated in the form of heat, noise and other waste. The environment is not just one of the three arenas, it is the essence of life and holds within its palm our fate as a species.

The social arena consists of our interactions, our health, social justice, human rights, well-being, social capital, community, diversity, culture and our built environment. Democracy, participation, privacy and social order are all important, as is equality and a sense of sovereignty and empowerment. Social cohesion is another significant element. Safety at work and play and the nurturing, education and appropriate development of our children sets in place '*society future*' (Dempsey et al., 2011).

The economic arena addresses economic freedom, equitable distribution of income, economic growth, balanced trade, economic efficiency, full employment, economic security, price level stability, and economic freedom. Emphasis on each goal may differ depending on the political philosophy that dominates in any given state.

Once upon a time (a very good place to start), humans were completely dependent on the environment. Indeed, we were highly skilled in understanding all the nuances and patterns of the natural world. We had to be, because our lives depended on ecological intelligence. As hunter gatherers, we had to work within the context of the landscape, just like every other organism on the planet. Each season brought its challenges, perhaps requiring migration to find nutrition. Feedback was important, meaning that interpreting the conversation around us was essential. Populations were kept in control by constraints such as the availability of resources and the occasional devastating wildfire or drought.

Many of the early civilizations formed their cultural and religious beliefs around Nature, and it was common for the deities to be half-human, half-animal. Think, for example of many of the Ancient Egyptian gods. Ra, the Sun god, had a hawk's head. His daughter, Sekhmet, the goddess of war, had a lion's head. Anubis, the God of the dead, had a jackal's head. Indeed, the Ancient Egyptians believed their gods could shape shift between animal and human forms.

The Ancient Greeks had lots of hybrid gods too. Pan, the flute-playing god of the wild, was half-man, half-goat. Centaurs were half-man, half-horse. Perhaps the earliest hybrid god was found in the Lascaux caves, dating back between fifteen thousand and seventeen thousand years, where a human with a bird's head is found on one of the cave paintings, together with a bison and a rhinoceros, which are certainly no longer native in Southern France.

The power of Nature and the spiritual interpretation of the landscape can be seen in present day hunter gatherer civilizations, and among the remnants of ancient indigenous peoples, such as the North American Indians and the Inuit. A deep relationship exists, and Nature is not only feared but embraced as one with society. The cultural intertwining of humans within Nature lies at the heart of these people, and society functions within this context, where the community is at one with its landscape.

Interestingly in modern times, a new movement, but with foundations in ancient indigenous values, called *buenvivir*, or 'living in a good way', has developed in the Andean nations of South America, giving us an insight into these earlier times (Escobar, 2011; Gudynas, 2011; Kothari et al., 2014). In Ecuador, it is inspired by *sumakkawsay*, the beliefs of the indigenous Quechua people of the Andes, while in Bolivia it is based on *sumaqamana* of the Aymara people. *Buenvivir* is now part of the Ecuadorian constitution since 2008, and of the Bolivian constitution since 2009.

It is an irony that one of the main drives for change in Bolivia is related to the mining of lithium, which has increased significantly in recent years since the rapid expansion of electric automobiles and mobile phones. Lithium is an essential component in rechargeable batteries. Some fifty percent of the world's lithium supply is located in the Salar de Uyuni region of Bolivia, the world's largest salt flats. Indigenous Aymara people have claims to the area. The drive towards the reduction in greenhouse gases, as promised by electric cars, is leading to significant tensions between the multinational companies desperate to profit from the lithium and the indigenous Bolivians. These tensions have contributed to the adoption of the *buenvivir* philosophy in their constitution, as the nation rediscovers its relationship between society and the landscape.

In 1968, Garrett Hardin wrote a paper called *The Tragedy of the Commons* (Hardin, 1968). In it he described a field, or commons, used by a number of herdsmen for grazing. To begin with there were just a few animals, and plenty of grass for everyone. However, with each year, the average herd size of each herdsman grew, and soon there was not enough grass, leading to starvation and the collapse of the herds. Underpinning Hardin's theory was the individual greed of each herdsman, who selfishly would seek to expand their herd.

While some have tried to argue that the Medieval commons on which Hardin's example was based were, in fact, self-regulating and well run (see Cox, 1985, for example), the tragedy of the commons plays out still across the globe, with over-grazing and over-exploitation contributing to soil erosion, habitat destruction and species collapse. If you need a clear example, try the Newfoundland cod collapse for size. Michael Harris' book, *Lament for an Ocean: The Collapse of the Atlantic Cod Fishery: A True Crime Story* (Harris, 1999), is well worth a read.

But in *buenvivir* and similar indigenous systems, this situation does not arise because the basic unit is not the individual, as it is in capitalism, but rather, the community and the environment. Here, the individual is relational, defined by their interconnectedness to others, socially and environmentally. We see this in many so-called 'primitive' or 'developing' societies, where leadership is shared among the elders, and decisions are made for the good of the community.

As a result of the *buenvivir* philosophy, the Bolivian government mines a modest amount of lithium each year, following the 'less is best' approach, rather than allowing multinational giants to carry out exhaustive mining. This is an example of avoiding the tragedy of the commons, by focusing on community rather than individual wealth, while protecting the environment. The lithium isn't going anywhere, and so a gentle approach will pay over many years, while being sustainable at both the social and environmental level. And this is because this philosophy is based on ancient belief, not the modern, individualistic philosophy that dominates Western society and delivers the commons tragedy of Hardin.

The *buenvivir* movement circles around the concept of living in such a way as to benefit both society and the environment. By doing this, the individual will then realize their place in the greater context. This is ecological and community empowerment, not individual empowerment. The key is harmony, both with each other and with our environment. It is a post-capitalist movement. Whilst capitalism stresses the right of the individual to own and to buy and sell, in *buenvivir*, these rights are secondary to the rights of the environment and the communities within them. Central concepts include sustainable degrowth, environmental accounting, austerity (i.e. using less of the Earth's resources) and more local production.

Buenvivir is also a post-development movement, meaning that it refutes the Western model of finance-based solutions. Pouring huge amounts of money at the so-called developing world, has, it is argued, only resulted in increased inequality, depletion of natural resources and the destruction of ecosystem services such as clean water and air (Dent and Peters, 2019).

This thinking is echoed in Africa. As Jean-Marc Ela, the Cameroonian sociologist, explains "*Africa is not against development. It dreams of other things than the expansion of a culture of death or an alienating modernity that destroys the fundamental values so dear to Africans…Africa sees further than an all-embracing world of material things and the dictatorship of the here and now, that insists on trying to persuade us that the only valid motto is 'I sell therefore I am'. In a world often devoid of meaning, Africa is a reminder that there are other ways of being*" (Ela, 1998).

Central to all of this thinking is a society where diversity and sovereignty lie at the community level, and that community is at one with the landscape. It sounds radical but it is actually how we were for ninety nine percent of our history as a species on the planet.

Edward Goldsmith, who founded the British Green Party, and wrote *Blueprint for Survival*, claimed that "*the problems facing the world today can only be solved by restoring the functioning of those natural systems which once satisfied our needs, i.e. by dully exploiting those incomparable resources which are individual people, families, communities and ecosystems, which together make up the biosphere or real world*" (Goldsmith, 1988).

But something changed around twelve thousand years ago. The last ice age had ended, and the warmer weather made life much easier. Two significant events occurred at this time: agriculture and permanent dwellings. Which came first is a point of debate, but it is clear that they occurred around the same time. By farming, it was possible to distance oneself from the power of Nature. Instead of foraging and hunting, you could grow crops and herd livestock.

This also allowed a division of labour, in that now, a part of the population could manage food supply, while another part could do other things, such as form an army. An army was useful as it allowed protection of your land, and expansion into someone else's land. Empire was born, probably when one settled tribe killed their neighbours and stole their farm. Soon excess food was produced and could be traded. Bartering and then currency followed, and a flourishing civilization, with artists, builders, poets and politicians, emerged. Economics was born.

The agricultural revolution and the accompanying urban revolution had many impacts, particularly in terms of the relationship between humans and Nature. By settling in villages and towns, and by farming, there was now a growing gap between the environment and society, and a growing relationship between society and economics. These trends would continue to the present day, interrupted only by the occasional natural disaster, a war or an outbreak of plague.

As the relationship with our landscapes weakened, we lost much of our natural intelligence, finely honed to allow us to survive in the wild. Instead, with increasing division of labour, a group of thinkers, inventors and technologists arose and had the time to think, invent and technologize. Information was now recorded in writing, and the storytelling culture of cave paintings and oral narrative would gradually be replaced.

Manufacturing gathered pace. Empires grew, utilizing the resources of newly conquered regions. Indeed, resource acquisition drove empire expansion. For example, it is thought that one of the main reasons that the Romans invaded Britain was to capture the tin mines of Cornwall. These mines contained most of the tin in the known world at the time (Levy, 2009). Tin was essential for the production of bronze and of pewter.

Indeed, the transformation that occurred at this time would set in place everything that we know today, from cities to farms, industry to running water, warfare to international trade, information technology (in the form of writing) and wealth inequality. As humans began to herd animals, diseases and parasites crossing between animals and humans also increased such as pox, bird flu, tapeworms, sleeping sickness and anthrax. Salt and sugar content of food increased leading to increasingly serious health issues such as stroke and type 2 diabetes.

The story of our journey from this point is really one of intensification and optimization, rather than invention. The industrial revolution merely increased the efficiency of the processes developed in the agricultural and urban revolutions, and economics became the central player. Now society and the environment were submissive to the economic god. *Homo economicus* was born.

And so, the ascent and expansion of humans on the planet is not really a complex tale at all. We settled, we farmed, we plundered, we traded, we invaded. And we kept doing these things, more and more intensely. But, as with the tragedy of the commons, every dog has its day. Resources are finite and Nature is vulnerable. And this is a problem, because we rely on Nature for something very important: *life support.*

When we look out of the window of our car, driving along a typical country road, you don't get the impression that the green stuff and the wet stuff are any big deal. As for the occasional bird, or the insect now smeared across your window, there's nothing particularly that would indicate that these were our intensive care nurses and doctors, whose lives, along with all the other lives on our planet, were part of a much bigger play, providing rainfall, energy flow and oxygen for us to breathe. The rain can feel like a nuisance, ruining a family picnic. The seasons bring their irritations. Too hot and humid in the summer, too cold in the winter, too many insects or that cursed dawn chorus of our avian friends, awakening us at some ungodly hour in the morning.

Yet when you actually write these things down in a paragraph like I have just done, you start to realize just how far we've wandered off-piste. The functioning of the Earth system is not there for us, but we are here because of the Earth system. We have given a name to the useful things that Nature does for us: *ecosystem services*. Ecosystem services are the activities of the biosphere upon which humans rely. Traditionally they are broken into four parts:

- supporting services (key services that underpin all the other services, such as photosynthesis, soil formation and pollination);
- regulating services (for example, the water cycle, the carbon cycle and the oxygen cycle that determine atmospheric and oceanic composition and circulation);
- provisioning services (for example, food provision, medicine provision, raw materials and wind and water energy)
- and cultural services (things like hiking, water sports, diving and educational trips).

However, the term itself is problematic. It implies that Nature is our servant, there for the sole purpose of providing us with service. *"What can I get for you sir?"*. *"Have a nice day, Madam"*. *"We're here to serve"*. But, really, it isn't like this at all. The Earth system has been doing its thing for over three billion years. It is not our servant. But the conditions it maintains are perfect for a warm-blooded, multicellular omnivore like ourselves, provided we don't tip it into a different state of mind.

Because the Earth system has changed its behaviour many times before. Changes in conditions can alter its functioning. Some of these alterations can transform the conditions on the planet to such a degree that huge numbers of species go extinct. The Earth system is not some little box of tricks you can put in your pocket. It's a mighty superpower, with vast resources and energy under its control. And when it tips, it really tips.

But surely everything's fine? The insects keep getting squished on our windscreens and the birds keep singing. The benevolent old Earth system just keeps bringing out the coffee and cookies and all is fine with the world, right? Let's drink some of that coffee and waken up. We've been gorging and feasting in the restaurant for far too long, oblivious of the expenditure. But the waiter has just appeared with the check. So has your accountant, and he's not looking happy at all. It's time to take a serious look at the cost of our excessive lifestyle.

VI.2. The five clear road signs that point towards criticality

The first of the big five road signs is, literally, the ground beneath or feet. Perhaps the most under-rated threat to our species is *soil erosion*. When I first started working in rainforests, I was struck by the incredible productivity of the system. Huge trees towered seemingly to the sky, with a vast diversity of life tightly intertwined. A single hectare of rainforest can have forty-two thousand different species of insect. A single tree can have forty species of ants alone. That single tree can store thirty tonnes of carbon, and release over three hundred litres of fresh water into the atmosphere. Two-thirds of the plants and animals on the planet live in these natural cathedrals.

Yet this whole amazing ecosystem relies on a layer of soil that is just six inches deep, on average. This life support system stretches across some fifty million square miles of land. Plants need soil, and animals need plants. Without soil, agriculture would collapse. As early as 9 000 BC, at the outset of the agricultural revolution that defined the beginning of modern civilization, soil tillage tools were in use in Jarmo in Northern Iraq (Troeh et al., 2004). The Ancient Greeks recognized the importance of soil, and included it as one of the four basic elements of life, the others being fire, water and air. Plato, around 400 BC, referred to soil as the mother.

But soil, like skin, is vulnerable to injury. Plato lamented that the soil of Greece was, by his own time, eroding away, observing that "*what now remains compared with what then existed is like the skeleton of a sick man, all the fat and soft earth having wasted away, and only the bare framework of the land being left*" (in Glacken, 1976). Around 60 BC, Lucretius, the philosopher and poet, recognized the seriousness of soil exhaustion in Italy. He thought that the Earth itself was dying (Green, 1942).

In the last one hundred and fifty years, we have lost fifty percent of the planet's topsoil through soil erosion. Lester W. Brown, the President of the Earth Policy Institute, puts it this way: *Civilization can survive the loss of its oil reserves, but it cannot survive the loss of its soil reserves*" (Brown, 2011).

So where has all the soil gone, as Pete Seeger should have mournfully lamented (as without soil, there aren't any flowers)? Quite simply, it's been washed away. Over-grazing by herds of cattle, goats and sheep have left fields bare and exposed to soil loss. Intense agriculture has had the same impact. Deforestation is one of the major issues here. Without the trees to bind the soil together, it simply erodes away. Soil takes many years to form, and losses far exceed formation. In China, for example, soil is being lost fifty-four times faster than it is being formed. This is causing a loss in economic and social terms too. For China alone, soil loss accounts for forty-two billion dollars per year, and impacts one hundred and seventy million people.

When the soil is washed away it ends up in the streams, rivers and lakes, causing huge disruption, as we shall see shortly. The soil can also block irrigation systems. This can topple empires. It is thought that the mighty Babylonian and Sumerian kingdoms collapsed due to soil erosion, blocking their irrigation systems (Hillel, 1991). Once the soil is gone, the risk of flooding after heavy rain increases dramatically, as there is no longer the water storage formerly provided by the soil.

It isn't just water erosion that removes the soil. Wind erosion is equally serious. One of the defining events of the twentieth century for the USA was the Dust Bowl, a ten-year collapse in agriculture caused by soil erosion from agricultural mismanagement in the 1930s. On 14th April, 1935, later called Black Sunday because the sunlight was blocked out by the dust, a huge storm removed some three million tonnes of topsoil from the Midwest. Around two and a half million people were forced to move, mostly to California, their struggle forming the inspiration for the classic American novel, *The Grapes of Wrath*, by John Steinbeck, written in 1939.

This book had a massive impact on me as a small schoolboy growing up in Ireland, bringing home to me the power of Nature and the dangers of abusing it. Ireland had itself suffered mass death and immigration following the potato famine in the nineteenth century. In Steinbeck's tale, The Joad family had been forced to migrate west, through appalling conditions, near starvation and penniless. The only hope was that the daughter, Rose of Sharon, was pregnant.

Near the end of the book, Rose of Sharon gives birth, but the baby is dead. All hope is finally extinguished. Uncle John is tasked with burying the still-born child of Rose. The baby was placed in an apple box. He carried it to the river. Steinbeck writes: *"For a time he stood watching it swirl by, leaving its yellow foam among the willow stems. He held the apple box against his chest. And then he leaned over and set the box in the stream and steadied it with his hand. He said fiercely, "Go down an' tell 'em. Go down in the street an' rot an' tell 'em that way. That's the way you can talk. Don' even know if you was a boy or a girl. Ain't gonna find out. Go on down now, an' lay in the street. Maybe they'll know then"* (Steinbeck, 1939).

As a result of farming intensification following World War I, and a severe drought, the Dust Bowl of the Thirties was particularly severe. While some argue as to the balance of the blame (see Cunfer, 2008, as an example), and that Kansas, rather than Oklahoma (the home of the fictional Joad family) was the worst hit area (Windschuttle, 2002), there is no doubt that a tragedy of the commons exacerbated the devastation. This extended the economic depression of the 1920s, and had worldwide implications. It has even been suggested that the Dust Bowl contributed to increased poverty in Germany, by extending the Great Depression, leading to a power vacuum that would be filled by Adolf Hitler.

The second major calamity is *soil salinization*, particularly in semi-arid areas. Here, large-scale irrigation is a necessity for agriculture. Vast amounts of fresh water, containing miniscule quantities of salt, are sprayed on the fields. As the water evaporates in the intense heat, it leaves tiny amounts of salt behind. Over time this salt builds up, creating irreversible damage.

Saline soils have only half the productivity of normal soils. Fifty percent of cultivated land is already affected in Africa (Ceuppens and Wopereis, 1999). Twenty five percent of European Mediterranean soils are already salinized. In Australia, soil salinization is the most pressing environmental issue, affecting huge swathes of agricultural land (Dehaan and Taylor, 2002). By 2050, it is estimated that fifty percent of the planet's crop land will be impacted (Bartels and Sunkar, 2005).

Soil salinization is a very serious problem, threatening the food supply to humans. It also creates problems for farm machinery, gas pipelines, railway lines and concrete structures, all of which suffer increased erosion. It is estimated that crop losses due to salinity amount to twenty-seven billion dollars annually. Combining soil erosion with soil salinity, the very future of agriculture is facing an existential threat. Given that populations continue to rise and that more crop land is being converted to produce green fuels, this poses a significant danger to humanity.

The third major calamity, and of our own making yet again, is *eutrophication*. Eutrophication is the loading of fresh and sea water with nutrients such as phosphorus, potassium and nitrogen. These three elements are the main agricultural fertilizers, loaded onto soils around the world in order to increase agricultural yield. These fertilizers are water-soluble, so they can move freely in the soil and reach the roots of the plants that the farmer is growing. However, as the water in which they are dissolved runs through the soil and drains into neighbouring streams, the nutrients are also washed out.

It is estimated that between thirty and eighty percent of applied nitrogen is lost to the environment, depending on the soil (Conway and Pretty, 1991). Once in the freshwater these nutrients have a catastrophic impact on aquatic life. Algae, enriched by the nutrients, begin to reproduce rapidly, as do cyanobacteria. The algae then die, and bacterial populations thrive on the dead algae, absorbing much of the oxygen in the water. As a result, fish and other aquatic animals die in huge numbers, suffocated by the lack of oxygen. Meanwhile the cyanobacteria produce toxins, killing more fish and posing a health hazard to humans and pets who come into contact with the poisons. Even buffalo can succumb to these toxins.

The nutrients next reach the sea, where the same thing happens. Large areas of our coastal waters turn into 'dead zones'. Dead zones are areas where fisheries collapse due to the lack of oxygen and the cyanobacterial poisons. The decline in commercial fisheries also has a huge impact on fish-eating birds such as the puffin, guillemot and kittiwake. Ironically, fish farms load the water with more nutrients to encourage fish growth, and this has a similar, devastating impact. Property values also drop as there is a terrible odor from the death and ecological mayhem. Drinking water can be tainted. The pH of the water rises, leading to fish becoming unable to detect crucial chemical signals that they need for their survival. In effect, they are blinded. In the US alone, eutrophication costs around two billion dollars each year (Dodds et al., 2009).

Soil erosion also contributes hugely to the problem, as the soil is washed into the waterways and out to the sea. I'm sure we have all seen our local rivers after a significant rain event, with the waters turned muddy brown because of all of the soil. Soil acts as a huge nutrient bomb when released into our waterways, with all the concomitant problems. In Europe, Asia and North America, fifty percent of freshwater is now eutrophic.

With so much disturbance in soil, water, temperature and habitat, it is surely obvious that the organisms who share this planet with us are going to suffer. And suffer they are. The *collapse of species diversity* is the fourth great calamity, and not just for the species involved, but for the human race and our prospects of survival. Extinction rates are currently running at between one thousand and ten thousand times background rates. Background rates are calculated from the fossil record, as the average extinction rate outside mass extinction events. It is estimated that one quarter of all species will be extinct by 2050, while eighty-eight percent of commercial fish and shellfish fisheries will have collapsed by 2050 if we do not alter current practice (Worm, 2016). In addition to over-fishing, dead zones from eutrophication have contributed hugely to this.

The impact of losing species is profound. Ecosystems are generally resilient things, able to cope with a myriad of stresses and assaults. This resilience relies on something called *species redundancy*. What this means is that there is more than one species for each job in an ecosystem.

Imagine two lollipop factories on a small island in the tropics. It is always sunny, and so each factory is outdoors. In each factory, there are three jobs: making the sticks, making the candy and putting the candy onto the stick. In the first factory, there is one person in each role, and each of the employees wears a red hat. So, there is no diversity. In the second factory, three people do each job, and each of the three employees at each station has a different colour of hat: red, blue or black.

Now imagine that the island had a dark secret, a huge predatory bird called the giant polka-dotted turkey vulture. This terrible bird-of-prey had a very specialized diet, humans wearing red hats. The vulture swept through one morning, eating all of the red-hatted humans. The first factory was decimated and all of its staff were eaten. The second factory, while mourning the loss of their beautiful, generous, warm-hearted and industrious red-hatted work colleagues, could still continue to function due to its greater diversity. Diversity brings resilience.

Ecosystems usually also have more species than roles. In fact, a healthy ecosystem will have between four and five species for each function. If one of those species goes extinct, the ecosystem can still continue to operate. However as further species are removed, the risks increase. Many of the world's ecosystems now have very little redundancy, and so they have lost their resilience. An old rhyme, *For want of a Nail*, reminds of how the small, often ignored entity can play a central role in a system, and if lost, can lead to its collapse:

"For want of a nail the shoe was lost;
For want of a shoe the horse was lost;
For want of a horse the battle was lost;
For the failure of battle the kingdom was lost—
All for the want of a horse-shoe nail."

Of course, we are losing species at present, but the ecosystem services are still, in general, being provided. So, what's the problem? The birds still sing, the bees still buzz and you just saw a butterfly sail by. Crisis? What crisis? The problem is that gradually, the redundancy in the system, so essential for its resilience, is being ground down. And suddenly, a tipping point is reached when the last species in a particular role disappears. Then, the entire ecosystem crashes, as a link in the functional chain snaps. Species redundancy can give us a false sense of security, but every species has a role to play, and their loss is significant, however invisible it may appear to be. As John Donne, the English metaphysical poet of the early seventeenth century, wrote, in his *Devotions Upon Emergent Occasions, Meditation 17*:

"No man is an island,
entire of itself;
every man is a piece of the continent,
a part of the main;
if a clod be washed away by the sea,
Europe is the less,
as well as if a promontory were,
as well as if a manor of thy friend's or of thine own were;
any man's death diminishes me,
because I am involved in mankind,
and therefore never send to know for whom the bell tolls;
it tolls for thee."

(Donne, 1839 [originally 1624])

Donne grasped the importance of Nature as a system, as did the great geologist, James Hutton, who, in 1785, is attributed to having said "*I consider the Earth to be a superorganism and...its proper study should be by physiology*"(in Lovelock, 1991).

This unified approach to the biosphere means that we are not dealing with individual extinctions, but the threat to the living planet itself. And if the system is threatened, then so is everything within the system. For despite referring to the three arenas in terms of our activities, there really is only one arena, the Earth system. Every species has a society and an economy, and they all rely on the environment. All of the biosphere's members stand on top of its delicate fabric, and rely entirely on it for their survival. The increasing loss of diversity is a shattering blow. The bell tolls for all of us when Nature groans. And that is why it is essential that we address the issues detailed in this section. We are at a crossroads, and sinking. But don't give up hope until you finish this book. Solutions are just around the corner, metaphysically speaking.

The fifth and final calamity is *atmospheric pollution*. What we put into the atmosphere affects us in different ways. Air pollution, particularly from fossil fuels and diesel emissions, reduces life expectancy by 1.8 years, making it worse than smoking (1.6 years), drugs (11 months), HIV (4 months) or conflict and terrorism (22 days). This makes air pollution the greatest global threat to human life (Greenstone and Fan, 2018).

Five and a half billion people, that is, seventy five percent of the world's population, live in areas where air pollution exceeds World Health Organization standards. This is nothing to do with climate destabilization. It is the consequence of dust, soot and smoke, from burning fossil fuels and biofuels. The impacts are huge, increasing the risk of asthma, heart attacks, stroke, dementia, infant mortality and reduced cognitive function. This is serious stuff, and needs urgent action. Climate change deniers have undoubtedly contributed to a lethargic attitude towards fossil fuels. However, particulate pollution is a reality that is a consequence of both fossil fuel and biofuel combustion and needs urgent attention at an international level.

The reality is that while a person can decide to stop smoking or drinking alcohol, thus improving their outcomes, air pollution is beyond the control of the individual, and requires global action. We all need to breathe air, and if the air is poisoned, we all suffer. It's passive smoking, but on a planetary scale.

Greenhouse gases in the atmosphere are rising. It is beyond question. The Keeling curve, recording the concentration of carbon dioxide in the atmosphere at the top of the Muana Lua Observatory in Hawaii since 1958, and named after its inventor, Charles Keeling, shows that CO_2 has risen steadily from three hundred and fifteen part per million in 1958 to four hundred and ten parts per million in 2018.

We have also known for over a century that CO_2 leads to increased air temperatures. In 1896, Svante Arrhenius, developed his hothouse theory, and in 1905, predicted that global temperatures would rise as CO_2 levels increased. His predictions match current warming very accurately.

Some have argued that it's just natural warming due to the interglacial (warmer period between ice ages) reaching its peak. However, Feng et al. (2015) showed clearly that carbon dioxide levels fell dramatically during the global recession of 2008, reflecting a dramatic drop in economic activity, particularly in terms of cement production and use, as building projects ground to a halt. This clearly points to human activity as the major contributor to carbon dioxide level change, and, given the fully proven hothouse impact, demonstrates that it is anthropogenic activity that has led to climate destabilization.

Indeed, more generally, scientific research clearly indicates that it is the activities of humans that represent the driving force behind the transformation of our ecosystems (USNRC, 2010). Together with the huge increase in damage from rising levels of particulate matter, fossil fuels and biofuels cannot be ignored. It's simple and undeniable.

Another set of atmospheric pollutants that have caused extreme damage are the chlorofluorocarbons (CFCs). Not toxic in themselves, CFCs damage ozone production in the atmosphere. Since ozone absorbs harmful UV radiation from the Sun, the damage to the ozone layer has led to a huge increase in skin cancer within human and animal populations, particularly near the poles. This is because ozone production is slower in colder regions. As a result, ozone holes have opened up in the northern and southern polar regions.

At least two in every three Australians will suffer from skin cancer by the age of seventy, and two thousand Australians die from it every year. It's worth reading this last sentence again, slowly. Antarctic sea urchin embryos, essential parts of the delicate food web in these cold waters, are damaged by UV. Normally they would have been protected by sea ice from the worst of the radiation, but with the melting of the sea ice, their numbers are collapsing (Lister et al., 2010). Coral reefs are also impacted, as are phytoplankton, the foundations of all marine ecosystems (Häder, 1997; Banaszak and Lesser, 2009). UV also impacts land plants, with reduced growth and reduced seed production (Jansen et al., 1998).

The atmospheric concentration of trichlorofluoromethane (CFC-11), a powerful ozone-depleting gas, had been declining since the implementation of the Montreal Protocol on Substances that Deplete the Ozone Layer, hailed as the first universally ratified treaty in UN history. However, it has been shown that depletion rates are slowing (Montzka et al., 2018). It has recently been demonstrated that the rate has slowed due to a one hundred and ten percent increase in CFC-11 emissions in Eastern China (Shandong and Hebei provinces) from new production (Rigby et al., 2019). When will we ever learn?

Increasing volumes of ice from glaciers and high mountains are melting. The resultant fresh water flows into the oceans, diluting the salt. This impacts upon ocean circulation, with huge effects on climate. Some twelve thousand five hundred years ago, a huge body of water formed from melting ice after the last ice age, called Lake Agassiz, burst its banks and diluted the North Atlantic Current and slowed it down. The North Atlantic Current carries warm water from the Caribbean to the British Isles. Indeed, in winter, half the heat in the UK comes from this current, and half from the Sun. The circulation relies on increasing salinity as the current crosses the Atlantic, through evaporation of water. At the British end, the heavy, salty water sinks and this is replaced by an inflow of water from the Caribbean. It's a bit like a conveyor belt.

But when Lake Agassiz burst its banks and flowed into the Atlantic, this diluted the seawater so much that the conveyer belt switched off. And this plunged Britain into another ice age within seventy-five years, called the Younger Dryas, lasting for one thousand years. This was a relatively local event. For example, during this period, Antarctic temperatures didn't really change. It was because of Lake Agassiz and its impact on the North Atlantic Current that only the Northern Hemisphere experienced the cooling.

With increased melting of glaciers currently, due to rising temperatures, there is concern that a similar event could happen again. And this is why the concept of global warming is a little misleading. Rather, it should be called *climate destabilization*. Some parts of the planet will heat up, but others may get colder. Some will enter into drought while others will be flooded. Because of the complexities of ocean and atmospheric circulation, increases in the overall temperature of the planet will have many different impacts.

VI.3. Why ecological damage matters to us

So, we have seen that the geosphere, hydrosphere and atmosphere are all being impacted, with significant consequences for the living world. As a result of this, ecosystem functioning is being threatened, and this impacts upon humanity.

We have already noted the serious health risks, both from UV and from particulate matter. Agricultural failure, as a result of soil issues (salinization and erosion) and climate destabilization (droughts and floods) is an increasing threat. Fisheries collapse due to eutrophication and overfishing along with shellfish damage due to ocean acidification threaten the food supply.

As carbon dioxide levels continue to increase, the oceans are beginning to become acidified. The carbon dioxide dissolves in the water, forming carbonic acid. In one way this is useful, because it buffers the Earth system, preventing carbon dioxide levels from reaching even higher levels in the atmosphere. However, there is a very serious problem developing in our oceans. As the carbon dioxide dissolves, this releases hydrogen from the water, and some of the hydrogen combines with calcium. Once this happens, the calcium cannot be used by marine organisms. The increasing amounts of hydrogen lead to the oceans becoming more acidic.

Since the industrial revolution, oceans have become thirty times more acidic. Ocean acidification is known as climate change's evil twin. This has devastating effects on sea animals that use calcium carbonate for constructing their casings, including many coral species, clams, oysters and sea urchins. The shells start to dissolve, and cannot be rebuilt in these acidic conditions. It has been twenty million years since the oceans have been this acidic.

You might think "*Well that's OK then. Things worked out twenty million years ago, and so they'll probably be fine again*". However, there are two differences between now and then. Firstly, the change has happened really suddenly due to human activity, and so biology has no time to adjust. Secondly, there were no humans on the planet twenty million years ago, and so recovery was possible. Today, because of other anthropogenic scourges such as over-fishing and fertilizer pollution, recovery is all but impossible unless we address all of these issues immediately.

Eutrophication and salinization also threaten the water supply, as does excessive agricultural irrigation. Green fuels are taking up valuable land that had been used for crops. Rising sea levels threaten many low-lying areas of the world, including the Pacific islands and Bangladesh, along with cities such as Guangzhou, New Orleans, New York, Mumbai and Osaka. The consequences will be huge numbers of refugees with nowhere to go, creating social instability. Such instability tends to polarize politics, with extreme left-wing and right-wing populism thriving under these conditions. This leads to increases in nationalism and tribal behaviour, producing further social disintegration.

Other social issues relate to urbanization, particularly in societies in Asia, where traditionally there was filial piety, and families lived together and cared for their elders. Increasing urbanization and the need to move regularly for employment means that intergenerational family structure is weakened. Demographic changes also create issues, as populations become skewed towards older age groups, due to increased longevity.

The problem here is that not only are there not the younger family member to provide support for the increasingly older populations, but that there is an increasing burden on pensions. More people are needing pensions for longer, but there is a relatively smaller workforce paying into these pensions. Also, parents are having less children. This is good for the planet, but means a smaller workforce in the future, with less money going into pension funds. As AI replaces humans in the work force, this problem will only increase. Many countries are facing significant pension shortfalls, as people stay alive for many more years after retirement.

Diversity and resilience are also important in human societies as well as in Nature. Diversity is a strength for any society, bringing lots of different ideas and cultural creativity to a community. There is a significant and real concern that globalization will lead to increased social vulnerability. If everyone adopts the same social and economic framework, there will be no diversity. And technology delivers globalization.

This is particularly relevant to AI. Imagine a world where identical algorithms produce identical outcomes, and where humans are nudged towards a homogeneous singularity, where we all sing the same tune. Diversity diminishes, social resilience declines and vulnerability increases. Post-development advocates argue that current development programmes, delivering Western, Northern solutions across the planet, will, ultimately, lead to societal collapse.

Silvio Funtowicz and Jerome Ravetz, both philosophers of science, observed: "*For as yet only a few independent thinkers can imagine a 'development' that means anything other than the achievement of a consumer society; and that is unsustainable on ecological grounds. To encourage the world's poor to consider developing along less destructive lines than ourselves combines two further contradictions: one, the physical impossibility of an environmentally benign consumerist society; and the other, the sin of the rich preaching the virtues of poverty to the poor*" (Funtowicz and Ravetz, 1994).

The increased mobility of humans, both in terms of transport across the planet and in terms of ideas across the internet, will erode diversity, because diversity requires a degree of isolation. This is a very challenging subject. How can you have a multi-cultural society that is both integrated and diverse? In Scotland for example, where I now live, the 2011 census showed that only one percent of the population could speak Gaelic, down from six percent in the 1881 census. In 1755, it was twenty-three percent. The last King of Scotland who could speak Gaelic was James IV and he died in 1513. And language is important, representing much more than words. It ties a people to their ancient origins and their connection to the landscape.

Yet with trade, unifying of the kingdoms, increased mobility and internationalism, it is impossible to stop the incoming tide, even if your name is Canute. Isolation preserves diversity, or the individual units of diversity. But is isolation a good thing? Too much isolation, like tribalism, prevents access to new ideas, whereas too little isolation means that diversity is diluted.

The current economic model is also producing greater inequalities both in terms of empowerment and wealth. Globalization has decreased inequality *between* countries, but there has been an increase *within* countries (Bourguignon, 2017). In order to understand why the economic philosophy of the Enlightenment has failed to deliver equality, we need to go back to its architect, Adam Smith.

VI.4. Adam Smith and his invisible hand

Adam Smith was a leading philosopher and economist, who laid the foundations for the dominant economic model of our times. He almost didn't do any of this. As a three-year-old living in Fife in Scotland, he was kidnapped by gypsies and was rescued from a nearby forest. His biographer, John Rae, commented that "*[Smith] would have made, I fear, a poor gypsy*" (Rae, 1895).

A poor gypsy he maybe would have made, but he went on to help shape the world as we know it, at a time when philosophers really made an impact (ask Marie Antoinette if you need more evidence of this). Sadly, his legacy has been somewhat harmed by a misinterpretation of his work, whether intentional or otherwise, that has contributed, some would say considerably, to the crisis we now find ourselves in. However more of that later.

Smith wrote two major works: *The Theory of Moral Sentiments* (Smith, 1759) and *An Inquiry into the Nature and Causes of the Wealth of Nations* (Smith, 1778). To understand Smith's economic theory, contained within his second book, it is essential to understand his first, and much less well-known book.

Theory of Modern Sentiments, published in 1759, focused upon the topic of how and why we make the moral decisions that we do. This area of ethics is, as we have seen, at the heart of AI as well, and has occupied philosophers for centuries. Smith took the view that we arrived at our own moral position having observed the behaviour of those around us. These observations enlightened us and we found a resonance with them. His view differed starkly from many of the French philosophers at the time, such as Condorcet, who argued that we experienced sentiments within ourselves first and then recognized them in others.

And so, according to Smith, at the heart of our decision-making, lay empathy, or sympathy, between humans. Thus, the morality of society emerged from its cohesive interaction. For this to work, we had to be open and engaged. Mutual sympathy formed a bond, leading to the modulation of self-interest, while preserving harmony. Our moral judgements are responses to our recognition of how others resemble us. Importantly both Smith and David Hume, another leading Scottish philosopher of the time, espoused that those around us were not acting as mirrors, but rather they were a part of us, and we were a part of them.

Following his first book, it would be another seventeen years until he published his second book, The *Wealth of Nations*. Prior to Smith's work on economics, the general trend was to run your nation's economy in a protectionist style. A nation's wealth was judged as the amount of gold it had in its vaults. Therefore, it made no sense to import things, as this involved giving some of your gold away.

Smith instead demonstrated that it was economically better to trade with other nations, because although you bought things, you could also sell things. His next revolutionary thought was that trade should have complete freedom, and this approach became known as *laissez-faire* (literally, *leave to do*) economics. Free trade would work because of what Smith referred to as the *invisible hand*. That invisible hand was the virtuous self-interest from his first book, wherein the drive to look after one's own interests was tempered by the emergent societal morality that each individual shared in.

Smith wrote that the individual *"intends only his own gain, and he is in this, as in many other cases, led by an invisible hand to promote an end which has no part of his intention. Nor is it always the worse for society that it was no part of his intention. By pursuing his own interest, he frequently promotes that of the society more effectively than when he really intends to promote it"* (Smith, 1904 [originally 1776]).

Finally, he set out the idea that a society would grow best within an open, competitive market place. So social well-being was in resonance with economic freedom. Tensions existed. Sympathy and self-interest do not appear as natural bedfellows. Monopoly and free trade seem at odds with each other. But it was the invisible hand, an emergent property of a functioning society, that would moderate and temper our self-interest, shifting and nurdling things in the right direction. The original nudge. Flowing like an electrical current through the whole thing was sympathy for one another, and so a functioning economy would emerge from a functioning society. In turn the economy would strengthen society. Humanity would ride this escalator all the way to Enlightenment utopia.

Yet somehow the first of Smith's books became forgotten, and the modern capitalist economy now overflows with selfishness, wealth inequality and exploitation. Adam Smith would be horrified. There is no sympathy nor virtuous self-interest to be seen. Wealth disparity destroys the very fabric of society, and the invisible hand is dead and buried, replaced by the iron fists of greed and exploitation.

VI.5. Kuznets and his curve: How ninety five percent speculation led us badly astray

Simon Kuznets, the Belarussian Nobel-Prize winning economist, set out a theory that economic growth, the paradigm of Enlightenment progress, would ultimately deliver social equality. In 1955, he presented evidence that, with time, while economic inequality would initially increase, in the long term it would decrease. This became known as the *Kuznets curve*, an inverted 'U' shaped thing of beauty and comfort to any neo-liberal economist. Economic growth would deliver social sustainability, and all would be well.

By the 1990s, his theory was extended to the environment. A number of economists, including Shafik and Bandyopadhyay (1992) and Grossman and Krueger (1995) claimed that economic growth would not only save society, but save the environment as well. They suggested that environmental quality would initially decline, but then would improve with economic growth. This became known as the *Environmental Kuznets Curve*, and underpinned sustainable development theory into the current century.

Emphasis was placed on developing the economies of 'developing' nations, and inputting finance from the private sector. Ecological degradation was now correlated with poverty, not economic growth. Economics was off the hook.

This laid the basis for the Millennium Development Goals (MDGs), a UN programme that ran from 2000 to 2015. The MDGs represented a neo-liberal agenda, based on globalisation and private sector donor-centric solutions (Saith, 2007). A strong economy would deliver sustainable societies and environment. Wealth would also lead to an increasingly educated, rational public, who would then make the right decisions. Mol (1995) has argued that the only "*possible way out of the ecological crisis is by going further into the process of modernization*".

Thus, economic growth would deliver a society that could progress, and an environment that could thrive. This became known as the *Ecological Modernization Theory* (EMT). This was very different from what Adam Smith had set out. He started with a healthy functioning society, from which individuals would gain empathetic skills and act with virtuous self-interest. The invisible hand would then emerge from this functioning society and guide our economic activity. The EMT instead emphasised economic growth as the invisible hand acting upon society.

Increased environmental protection would come from an increasingly educated society, while maintaining economic growth. Furthermore, environmental capital can be replaced by man-made capital. So as long as you have capital, it doesn't matter what form it is in. This was the basis of the *weak sustainability* argument: that economic growth would deliver social and environmental sustainability and that you could substitute natural processes and products by man-made processes and products. Economics is neither dependent upon nor embedded in the environment. The environmental arena could just be ignored. It's the Economy, stupid.

However, both the Kuznets Curve and the Environmental Kuznets Curve (EKC) have come under significant attack more recently. David Stern (2003) states that "*The evidence…shows that the statistical analysis on which the environmental Kuznets' curve is based is not robust. There is little evidence for a common inverted U-shaped pathway that countries follow as their income rises*". But we don't need to look any further than the Nobel Laureate himself, who stated in 1955, that his research involved "*perhaps 5% empirical information and 95% speculation, some of it possibly tainted by wishful thinking*" (Kuznets, 1955). This is shaky ground on which to build a defense of the Enlightenment.

Milanovic et al. (2007) demonstrate that there was no increase in inequality from the Roman Empire to the 1880s, again contradicting Kuznets. Skene and Murray (2017) observe that as the global middle class rises from one billion to four billion by 2050, although pollution per unit manufactured goods may decline, the accompanying huge increase in demand will lead to much greater environmental damage. Neumayer (2002) has shown that people's revealed preferences still represent pollution-intensive material goods being highly valued.

The treadmill of production theory, established by Allan Schnaiberg (1980), contradicts the EMT, and clearly states that even with increased efficiencies in production, savings are outstripped by increases in the scale of production (Jevons, 2001; Ewing, 2017).

Recent work has pointed to a U-shaped curve (Dietz et al., 2012), or a similar N shaped curve (Fujii and Managi, 2013), wherein economic growth leads to increased environmental damage. In a detailed study, Jorgenson and Clark (2011) show that there is a clear and increasingly strong coupling between economic growth (as measured by GDP) and environmental damage over the last fifty years, and this is the same for developed and developing nations. This acts as a proof for the treadmill of production theory and rejects the ecological modernization theory.

Such results are shocking and extremely important. They undermine the main pillar of neo-capitalist development policy: that economic growth delivers better societies and environment. And this has led to a backlash, with post-development thinking, often based on pre-Enlightenment indigenous thinking, now challenging the major global organizations such as the World Bank and the United Nations.

The most recent international sustainability policy, *the Sustainable Development Goals* (SDGs), have turned away from the Kuznets model, where economic growth would deliver everything else, to an entirely different approach, embracing systems thinking (Skene and Malcolm, 2019). Even in its formation, the SDG policy differed hugely from the Millennium Development Goals (MDG) policy development, in that civic society, developing nations and grass roots movements were brought to the table from the outset, rather than a small group of bureaucrats in the UN headquarters, as was the case in the MDGs (Waage et al., 2011). The SDGs also address the lack of emphasis on social justice, equity, empowerment and human rights in the MDGs (Fukuda-Parr, 2010; Langford, 2010).

So, fundamentally, economic growth does not deliver environmental sustainability, and neither does efficiency in production. This needs to be placed at the top of any discussion of how to move forward. Also, with the EMT now in ruins, the idea that economic growth delivers social sustainability is also refuted. It will require honesty and frankness within the multinational businesses, governments and global organizations to face up to the implications of these findings.

However, as we have seen at the start of this section, this reality check is not an optional extra, but a fundamental necessity if we are to avoid a crash landing whose deadly flight path we are already committed to. Change must come, and the treadmill of production theory actually is an extremely helpful compass of where not to go, as we shall see later, particularly in terms of how AI can help pull us out of a terminal descent. Since the whole humanity is on this plane, it's time to have a serious word in the ears of our pilots, navigators and flight crew. By this I mean our elected officials.

Given that inequality grows with economic growth, this surely points to the need to prioritize social and environmental cohesion, and then to discover the economic solutions that can deliver these outcomes, rather than the other way around. Contained within the current economic model of growth is the unequal partitioning of resource sourcing, cheap labour and waste production. Less developed countries occupy a very different place in the world economy compared to more developed nations. The former act as a source for raw materials and a sink for the concomitant pollutants associated with their extraction, while the industrialized nations merely use the products of this pollution. Thus, there is an unequal burden of production, in terms of ecological and sociological damage, upon less developed nations (Jorgenson and Rice, 2012). It's frankly outrageous.

Hence, not only are the arenas of society and environment damaged by an increasing economic arena, but the effects are uneven, with the Western, Northern world transferring their burden onto the Southern, Eastern world. Of course, the consequences of such off-shoring of environmental and sociological damage, while often concentrated in these less developed nations, are equally often impacting on the planet as a whole. Also, many of the countries onto which we push the cost for our consumption upon are also important food producers for the globe, so our strategy, and strategy it clearly has been since the days of empire and beyond, will return to bite us.

So, we have seen that things are going badly wrong in the social arena. As a result, there is no invisible hand tempering the economic arena, which has scampered out of control. This has led to the economic arena consuming the environmental arena and abusing the social arena.

So where does AI come into all of this? Well in the next few chapters, we will examine the current role of AI in the three arenas, before considering the controversial topic of the relationship between technology and sustainability. And then we will set out a new and transformative role for AI: literally the salvation of our place within the Earth system.

SECTION VII

Forget the Romans. What has AI Ever Done for Us?

"Every person fills out quite a few forms in his life, and each form contains an uncounted number of questions. The answer of just one person to one question in one form is already a thread linking that person forever with the local center of the dossier department. Each person thus radiates hundreds of such threads, which all together, run into the millions. If these threads were visible, the heavens would be webbed with them, and if they had substance and resilience, the buses, street-cars and the people themselves would no longer be able to move."

–Aleksandr Solzhenitsyn, *The Cancer Ward* (Solzhenitsyn, 1968)

We see that trouble stirs across the three arenas (economy, society and environment) and that the dominant worldview is a distorted version of Smith's vision. Change is needed urgently. This is the biggest challenge ever faced by humankind, and is all of our own making. It will take an ecologically intelligent approach to resolve, and we'll detail this in the final section. But how can AI play a role here?

In this section we examine the current importance of AI in each of the arenas. The aim here is not to approve nor disapprove of these roles, but to examine how we currently use AI and then think about the implications of these uses, either in a business-as-usual scenario, where we keep on the path we are on, or the alternative route, where we steer around the gulch that awaits us on the former path. We then face up to a major issue: should we use technology to build a sustainable future at all? There are two ferociously opposed camps here: weak sustainability and strong sustainability. We take a wander around both camps before facing this issue head on: can technology deliver sustainability?

VII.1. AI and economics: The best of things or the worst of things?

AI is widely viewed as having a huge impact on economics. As to the nature of that impact, opinions vary widely. To misquote Dickens: *"It is the best of things, it is the worst of things"*. Mark Carney (2018), the Bank of England Governor, stated that *"Data is the new oil"*. His deputy, Ben Broadbent said that the economy could be awaiting its next big breakthrough, possibly as a result of AI (Isaac, 2018).

Yet hanging over all of this optimism is a fear that employment could be badly impacted. And paid employment forms the bridge between economics and society. Indeed, one could argue, employment is the bedrock upon which a modern society is built. Remember the images of stockbrokers throwing themselves off buildings in New York during the Great Depression of the 1920s? Work not only provides income, but it also provides meaning, status and a sense of identity. If AI does indeed lead to significant unemployment, what impact could this have on the mental well-being of a population now made redundant by machines?

Frey and Osborne (2017) claim that the advent of AI will mean that almost fifty percent of jobs in the US economy are at risk of being replaced by technology. KPMG International (KPMG, 2016) report that, by 2025, some seventy seven percent of Chinese jobs could be threatened and that some fifteen million jobs will be eliminated due to robotic automation in the UK alone.

But concerns do not only focus on job losses. Manyika et al. (2017) found that by 2030, as many as three hundred and seventy-five million people (fourteen percent of the global workforce) will be impacted by technological skill shortages worldwide and will need to retrain.

It is always difficult to predict the impact of new technologies. Jean Baptiste Say, the famous French economist, wrote, in 1828, about the possibility that cars might replace horses: "*Nevertheless no machine will ever be able to perform what even the worst horses can – the service of carrying people and goods through the bustle and throng of a great city*" (Say, 1828). He certainly got that one wrong. Similarly, we cannot truly predict the impact of a fully operational AI network on employment.

It is worth reflecting, too, that in the nineteenth century, seventy percent of Americans worked on farms. Since then, industrial technology roles, such as drillers, planters and combine harvesters, has replaced ninety nine percent of these roles (Kelly, 2012). So, it has happened before. And the mass production factories of Ford and others were able to absorb this redundant workforce. But where would all these people go today, when even mass production is automated?

In terms of AI replacing the human work force, Steven Pinker (2007) writes "*As the next generations of intelligent devices appear, it will be the stock market analysts and petrochemical engineers who are in danger of being replaced by machines. The gardeners, receptionists, and cooks are secure in their jobs for decades to come.*" This is summed up in Moravec's (1988) paradox, stating that while high level reasoning requires relatively little computation, even basic sensory-motor skills require huge amounts of computation.

What is likely is that AI will force what Brynjolfsson and McAfee (2011) refer to as a *Great Restructuring* upon us in terms of human employment, either as collaborative partners with technology or as competitors.

Arntz et al. (2017) argue that such doomsday scenarios are misplaced as heterogeneity within roles (i.e. a given job will involve a diverse range of tasks, not all of which are automatable) means that, according to their calculations, only nine percent of roles could be fully automated in the US. They further argue that roles will evolve to embrace technology in order to work together, just as has happened with other transformative technologies in the past. This is called *intelligence augmentation*. Here, the business represents a socio-technical system, and advances in AI can be matched by changes in human roles, wherein intuitive thinking comes from the human and analytical thinking from the AI. The augmentation of labour would allow humans to focus on high value tasks, while robots carry out low value tasks.

Another aspect that is often overlooked is that AI may provide literal, physical augmentations to humans, creating a semi-robotic human with increased strength or dexterity, but still connected to a human brain. Brynjolfsson and McAfee (2011) explore a number of these AI-human working relationships, including technological consistency married to human insight. This leads to a complementary working relationship between AI and humans, rather than a competitive, winner takes all relationship.

Acemoglu and Restrepo (2018) point out that through every technological revolution, new roles have always emerged. For example, from 1980 to 2007, total US employment grew by seventeen percent and half of this growth was accounted by new job titles. Another example is the introduction of ATMs in the USA, leading to an increase in bank clerk jobs, rather than a decrease, as money saved by banks in handling cash could be redistributed into opening new branches to compete with other banks, where the clerks could focus on more 'human' roles, such as discussing mortgages, loans and business plans.

AI promises to lower prices of products, saving on human labour and payment, while increasing efficiency and productivity. It has been argued that reduced costs would mean that negatives related to job losses might be offset, creating a more egalitarian world.

Certainly, automation of a workforce has many advantages. They could work twenty-four hours a day (one hundred and sixty-eight hours each week compared to forty hours by humans), would not need toilet breaks, holidays, parental leave nor sick days (other than for repairs). New software could be added, instantly increasing ability, rather than having to send humans to expensive training courses. No strikes, complaints nor wages. No pension fund, lack of consistency nor contractual negotiations, and no job interviews nor tribunals. As an employer, what's not to like?

AI is a general purpose technology (GPT), in much the same way as electricity, the internal combustion engine and steam were. These technologies not only transform the way we live, but also lead to a cascade of innovations flowing out in Cornucopian quantities. There is no reason to expect anything different from AI, particularly in terms of the economy.

Society interacts with the economy in three ways: employment, financial reward and produce. The environment interacts with the economy as a source and a sink of industrial activity. Thus, any great restructuring of the economy is bound to have significant implications for both society and the environment. However, many argue that while previous GPTs have certainly led to significant shifts in work (for example from agriculture to manufacturing), ultimately, more jobs were created by these technological revolutions than were lost.

Hence, there is an argument that any great restructuring as a consequence of the advent and spread of AI will follow a similar path. Here, AI will generate increased capital through increased efficiency, and this will produce new, non-automated work opportunities for humans, counter-balancing any displacement effect. One such area could be in sustainability roles. By this I mean employment could be created by the need to recycle material, repair damaged ecosystems and restore ecological integrity, all of which will be essential if we are to remain on a functioning planet.

A restorative revolution leading to an age of sustainability will be more fundamentally important than a technological revolution, though as we will shortly explore, technology could be the ultimate weapon in our fight for our very futures. Other new tasks that could emerge from AI directly were categorized in a recent Accenture report (Accenture, 2017) as falling into three categories: sustainers (ensuring that the technology continues to provide accountable, fair outcomes and that issues are addressed as they arise), trainers (training AI systems) and explainers (interpreting AI outputs for customers).

Of course, predictions of the future impact of AI tend to focus on what is possible in terms of capabilities, rather than what is required in terms of functioning, successful businesses. Giorgio Presidente (2017) commented: *"The current debate on 'the future of work' or 'jobs at risk of automation' seems to implicitly adopt a pure science-push view, which assumes a path for technology driven by what science makes achievable, rather than what is needed by firms."*

An interesting aspect of AI-driven changes in the economy relates to the potential political implications. Populist politics, be it right- or left-wing, often centres around employment and earnings. The 2016 US presidential election was impacted by fears of a strengthening Chinese economy damaging American manufacturing, climate change advocates damaging jobs in coal and oil, and concerns over immigration. Could AI have the potential to ignite a further polarizing, populist revolution? If the perception is that AI is displacing humans from work and that it represents an elite control of society, there could be a political backlash (Levy, 2018).

It has happened before. As technology began replacing textile jobs during the Industrial Revolution, a movement called *Luddism* rose in Nottingham, England, and between 1811 and 1816, textile machinery was destroyed and mills were burned. In 1830, agricultural workers in South East England destroyed threshing machines, in what was called the *Swing Riots*.

While the French Revolution had begun due to the spiralling cost of bread (remember Marie Antoinette's response to the starving peasants who could not afford bread: "let them eat cake"), by the early nineteenth century, the cost of corn had plummeted and the workers, dependent on land owners because of the removal of common land by a change in the law, saw their earnings collapse. Machinery was the final straw, as landowners and mill owners turned to technology to reduce costs.

Of course, the memories of the elite heading to the guillotine still remain with those in power today. This may yet prove the most important constraint on a completely unfettered AI system wrecking ruin in society through displacement of workers. Because workers rebel, topple governments and lead revolutions, other solutions may have to be found. It may be better to subsidise workers, even if they are not really required, than to risk the ire of a rebellion. Thus, there has been consideration of some sort of capital redistribution from the savings made by AI to the human victims of technology.

Economic growth delivered by AI may well benefit the elite, but it must be seen not only to benefit the poor, but to sustain a belief within the hearts of the poor that they could one day become rich. As Delaney (2005) states, *"The tolerance of the poor for the existence of the wealthier classes depends on a belief in economic growth."* If AI removes upwards of fifty percent of the jobs, then economic growth would cease for those workers, creating a dangerous underclass who would view technology as the sword of the elite and the creator of inequality. Of course, the elite may not want to redistribute their wealth to the poor, and may send in the robots to nullify any thought of revolution.

AI itself is now big business, with Ismail (2017) estimating that annual spending on AI by business will reach forty-seven billion dollars by 2020. The importance of AI in creating a digital economy has been recognized by governments. In July 2017, the G20 group of leading economic countries declared that the *"digital transformation is a driving force of global, innovative, inclusive and sustainable growth"* and undertook *"to foster favourable conditions for the development of the digital economy and recognise the need to ensure effective competition to foster investment and innovation"* (G20, 2017).

Of course, decreases in hardware costs are potentially not as feasible as reduced software costs, particularly when key chemicals needed for the hardware pass peak levels or are subject to cartels and geopolitical ambiguities. Take for example, the rare earth metals. China supplies over ninety percent of these metals, essential to modern technology. Neodymium is essential for tiny magnets that are key to the miniaturization of technology in, for example, cell phones. Without it, the typical cell phone would be the size of a shoe (I remember when they were that size). However, with one country controlling supply, that country can also control component manufacture and pricing.

China has, in the past, ceased supplying these metals to Japan due to a political dispute, leading to problems for the Japanese technology industry. Rare earth metals are again being used as a weapon in the ongoing trade war with the USA. With only a few nations supplying some of the essential material elements of technology, there is plenty of scope for the formation of cartels, much like with diamonds, that artificially maintain high prices, irrespective of resource availability. Hence, there is no guarantee of reducing costs in hardware.

There is also an environmental cost to mining rare earth metals (Ichihara and Harding, 1995). For example, the production of a single tonne of refined rare earth metals from Bayan Obo, in China, produces one and a half tonnes of radioactive waste, two hundred cubic metres of acidic water and sixty three thousand cubic metres of harmful gases, with many workers suffering from pneumoconiosis (black lung) and occupational poisoning from lead, mercury, benzene and phosphorus (Jiabao and Jie, 2009). In an AI world of the future, energy and natural resources will be the major constraints.

Another interesting impact of AI could be on international business. With AI replacing labour in factories, there will be fewer jobs in manufacturing, and so there will be little or no wage differences between nations. This means that there will no longer be an advantage in moving manufacturing plants to lower paid regions. Thus, there could be a "re-shoring" of manufacturing (Cowen, 2018). However, this would lead to a decrease in manufacturing in poorer countries. Africa could miss out on an industrial age (other than in raw resource mining), and move directly to a technological age.

There would also appear to be a productivity paradox thus far, wherein there has not been a concomitant increase in productivity with increased technological advance. Real income has stagnated since the late 1990s while non-economic measures, such as life-expectancy, have fallen (Case and Deaton, 2017).

The use of AI in branding and marketing is now considered to be extremely important. West et al. (2018) suggest that there are three main areas where AI can revolutionize the field:

1. Improvements in operational efficiency – leading to consistency in delivery, an essential component in building a solid brand;
2. The use of natural language processing (NLP) leading to improved customer care;
3. Machine learning, allowing targeting of specific customer profiles, personalizing the brand.

All of these make the brand your friend, rather than a door-to-door salesman jamming his foot in your door.

Ivanov and Webster (2017) raise an interconnected issue. As AI becomes embedded within our homes, it may well actually take over our consumption choices, by selecting and purchasing for its owner. Here, AI becomes the *de facto* customer. Amazon's Alexa can already order pizza from a well-known pizza chain. As a result, businesses will have to advertise their products to an AI consumer market, not to a human consumer market. This will require a huge rethink, as currently marketing is based on human psychology, targeting human desires, priorities and decision-making. Ivanov and Webster refer to this as *r-marketing*, marketing aimed at robot customers.

It would not be a huge leap in the dark to think of an autonomous vehicle arranging its own yearly service and taking itself to a booked appointment, handled by a robotic staff, that it had arranged while the human owner is at work or asleep. How would a cowboy car dealer cope trying to trick a robot into a questionable purchase of a stolen or damaged car?

Of course, the autonomous vehicle could be fitted with algorithms by the manufacturer that ensured it spent more of the owner's money, and at businesses owned or at least linked to the automobile producer, without the human owner being able to control this. In other words, the AI could be "brain-washed" and hypnotised by its designer to make the decisions that the manufacturer wants it to make, meaning that there would be no need for marketing at all! This would have huge implications for competition and monopolization.

As with so many related aspects of any new suite of technologies, much depends on the degree of trust and openness, and a lack of trust could create paranoia in terms of our view of AI. What is my home robot buying, why and under whose control? These types of questions underpin much of the concern over new technology, where big data, hidden algorithms and personal and national security issues will form significant challenges in terms of the use of technology in commerce.

In terms of business management, it will be necessary to extend human resources to include the robots employed, especially once they really start to learn and make decisions. Contractual issues could well arise. If true intelligence develops, and AI becomes a living entity in all but its chemistry, then sentience becomes a significant issue. In other words, robots could become emotionally connected, to each other and to the humans they serve, meaning that a code of ethics, support, counselling and contractual negotiation may need to be developed and 'difficult' robots may need to be chastised or face employment tribunals.

This needs careful consideration. Furthermore, assessing the return on investment (ROI) of AI is challenging, because rather than depreciating in value over time like, say, a machine that makes coffee cups, the AI will increase in value over time as it becomes more sophisticated through machine learning, enabling it to fulfil more advanced roles. Thus, asset *appreciation* actually comes into play. It may also be difficult to put a value on increasingly advanced functioning, as initially, we may not fully understand what any new capabilities can be used for.

The question remains as to if AI can become the idealized *Homo economicus*, completely rational and optimal in its decision-making. While Adam Smith presented an economic model that relied on human empathy and sympathy, the societal conscience or invisible hand, AI would need no such mentoring. Indeed, devoid of the emotional white noise, the passions of the heart and frailties of the mind, AI should be much better at rational thinking, surely? Yet the question remains as to whether or not machines can be truly rational.

Furthermore, as AI systems begin to interact with each other, how can diversity be maintained within the group of AIs while still optimizing for each AI? In societies, diversity is, qualitatively, an emergent characteristic. It cannot be programmed and is not goal-orientated. Thus, it is difficult to see how economy-focused AIs would have any motivation to diversify. Expertise brings specificity, and competition and selection always reduce diversity anyway.

Another issue arises in terms of the temporary nature of a given AI within an AI society. If programmed, it could adopt a different personality in order to gain an advantage, but if all of the other AI agents did this, there would be huge uncertainty, leading to unpredictable outputs. It would resemble the ancient Egyptian gods who could shape-shift between different forms. Forget the bull of Wall Street. One day you could be dealing with a panda, and the next a tiger.

Currently, most of the trades on the US stock market are carried out by automated trading algorithms (MacKenzie, 2018). And yet, surely, these algorithms will operate to maximize profit, as business plans have always done? Without a truly integrated system, where society and environment are meaningful voices at the table, AI-dominated economics could pose a very significant threat to humanity and planetary health, by perfecting the damaging economic practices of humankind.

Another economic issue arises here. If technology reduces costs and delivers improved purchasing power for humans, thus meaning that wealth re-distribution would not need to be so large (in that the cost of living declines), then deflation could result, plunging us into a depression that would destroy investment.

Korinek and Stiglitz (2017) explore this in detail. They reflect on the Great Depression of the 1920's, caused by automation in the agricultural sector, massively reducing employment in the sector, thus reducing the cost of production. This led to a decrease in income and a reduction in purchasing power for property and for urban goods. Both urban and rural sectors failed. In many ways, World War II, and its aftermath provided the stimulus to retrain rural workers for urban roles, recovering the economy.

AI has the potential not only to impact on production of goods, but also on the innovation process. Exploring new strategies, concepts and methods could all have consequences for the economy. If an AI singularity is reached, exponentially improving, the consequences for both production and idea generation innovation could lead to an explosion in economic growth. James Barrat (2013) suggests, in a book of the same title, that such will be the creative power of a fully developed AI system, that it could be *"our final invention"*.

Constraining singularity are a number of factors. Firstly, AI may not be able to achieve creative superintelligence, nor replace all facets of human labour necessary to deliver singularity in terms of production. Innovation potential may not be infinite. Furthermore, resources and waste may constrain production. For example, we may have superefficient AI controlling fishing, but there will be a limit to the number of fish in the sea. Also, consequences in production such as fish food, fish waste, available habitat, essential physical laws such as diffusion of oxygen, production of CO_2 from respiration and cycling of materials through the biogeochemical pools would all impose limits to production.

This brings to mind Liebig's law of the minimum, named after the German Chemist, Justus von Liebig, which states that yield is proportional to the amount of the most limiting nutrient, whichever nutrient it may be. Remember our lollipop factory? The process relies on the sugary sweet that sits on top of the lollipop stick, and the stick itself. The maximum production of lollipops will be limited by the most limiting content, be that sugar, the stick, or the flavouring.

Indeed, the market size for lollipops may also limit things, as may the distribution network, the storage warehouse availability and the number of lollipops that can be consumed. Increasing concerns over public health due to deaths by obesity may also create limits, shrinking the market, as customers perish. Fundamentally, computers don't eat lollipops. The sweet syrup messes with their motherboards. Human customers are needed for consumption. It is also worth noting that the limited market for paperclips is likely to reduce the likelihood of a paperclip apocalypse.

It is not expected that AI will impact on all sectors equally because not all sectors can be improved by AI equally. Aghion et al. (2017) argue that AI will lead to the expansion of firms who can benefit from external knowledge transfer, wherein they can gain access to solutions from elsewhere, whereas more traditional self-contained firms will not benefit as much. However, the counterargument is made, wherein firms utilizing knowledge transfer may also suffer in the reverse direction, through knowledge theft. Companies that are more stand-alone may thus benefit in the longer term. This can be called the *innovation/imitation paradox*, where free-flow of knowledge may intensify competition but reduce diversification. If a company spends resources on innovating new ideas and products, but these then diffuse throughout the system, that company may not make the financial gains to offset this.

Another interesting aspect of AI in terms of economics is that of the potential power inequality becoming technological as much as financial. The haves and the have-nots are traditionally separated by money. Money buys influence and influence buys power. The leaders of the computer technology world are the new rich, moving ahead of more traditional wealth such as manufacturing, fossil fuels and land. Indeed, fossil fuels, having reached peak levels and becoming unpopular within a more environmentally-aware society, are likely to diminish in terms of delivering wealth and power.

As AI systems begin to play significant roles in controlling energy supply, transport, financial services, health and security, this power inequality will increase. This poses its own set of problems, and leads to increased inequality. Advances in genetic engineering along with AI may provide the elite with unprecedented advantages over the poor, becoming a masterclass, referred to by Yuval Harari (2017) as "the gods" in his book *Homo deus*, while the rest of humankind become "the useless".

As Cassius, in *Julius Ceasar* by William Shakespeare, reflected, in Act I Scene II:

"Why, man, he doth bestride the narrow world
Like a Colossus, and we petty men
Walk under his huge legs and peep about
To find ourselves dishonorable graves."

In terms of power, the use of AI in military operations has the potential to become the most important factor in societal and economic terms. Horowitz (2018) points out that AI is more likely to be an enabler than a weapon, but could be targeted by weapons such as hacks. One factor that could determine the future trajectory of AI is if its development is driven by commercial or military interests.

If a duality exists, i.e. that military research is separate from commercial work, then it is likely that nations such as the USA and China will have advantages over poorer nations, in that they can put more resources into secret algorithms, giving them *'first mover'* advantages. However, if AI applications in the military are derived from commercial algorithms, then any country with a reasonable AI research programme will be able to develop military applications, thus diminishing potential power inequalities, and, therefore, wealth inequalities. Of course, this is a central argument for having a well-funded military AI research programme. This was the case in the early development of computers and AI.

The potential of AI in war ranges from controlling the targeting of bombs or torpedoes (already in use) through to algorithmic warfare, where entire wars are strategized by a general artificial intelligence, pitting its wits against another AI on the opposite side. A deadly and potentially existential game of chess could ensue. War is fundamentally an economic stimulant, and the scenario is not unimaginable where an economically-minded AI starts a war as part of its economic plan.

As we have already noted, many would claim that the Second World War provided a restructuring necessary to shake off the Great Depression. Others would argue that this represents what is called the broken window fallacy (Bastiat, 1964), wherein if a child breaks the window of a butcher's shop, the butcher must pay a glazier to fix the window. However, this doesn't result in a net economic gain, as he would have spent the same money on something else, perhaps investing it back into his business and making a greater profit. And so, breaking the window, a bit like war, doesn't have an overall benefit.

It could be counter-argued that in war, and during the aftermath, the child not only breaks the window but drives the butcher out, and takes over, gaining hugely from the action. He then destroys all the other butchers, raises his prices and builds a vast empire. Think oil gains from recent wars, or control of shipping routes.

Victory in war allowed Britain and other European nations to take mineral and human resources (slaves) from many African and South American countries and build their empires. It would be impossible to argue that the defeat of the indigenous North American Indians and the resulting economic powerhouse that is the USA did not lead to a huge increase in economic prosperity for the victors. And so, if AI can provide the edge in war, it is likely to deliver huge benefits to the winner. Furthermore, autonomous vehicles such as tanks and planes, already under development, will mean that countries with small populations can have a significant military might.

Military investment tends to be much more easily sanctioned by government than purely academic expenditure, especially in economically powerful nations. But just as initial advances in computing, atomic energy research and space exploration were all driven by military expenditure but then led to huge benefits for the commercial world, so military applications of AI will doubtless advance the field much more quickly than without military funding.

The fact that the governments of Russia, China and the USA are all recognizing the potential of AI in warfare, while disturbing, will likely lead to much faster development of the field. However, while atomic bombs came before nuclear power plants, commercial AI is already up and running, partly because the military withdrew funding in the 1970s following a perceived failure to deliver. Thus, commercial AI is already advanced, meaning that there is likely to be access to much of the knowledge already. This means that there is unlikely to be a huge qualitative advantage for large nations (i.e. everyone will have access to relevant algorithms).

Indeed, commercial companies are now pulling out of contracts involving government military projects. Recently Google withdrew from the Joint Enterprise Defense Infrastructure project at the Pentagon, which was working to utilize cloud computing to speed up decision making in a battlefield situation, and was worth ten billion dollars, because it "*couldn't be assured*" the Pentagon work "*would align with our AI principles*" (Waters, 2018).

Rich nations may return to quantitative advantages, as they have more money to build more robots. However, modern technology requires a large number of scarce resources (for example rare earth metals), which are unevenly distributed across the planet. Control of these could come into play, impacting on quantitative advantage. Finally, any wireless AI control system, relying on communication from large banks of computers remotely, is susceptible to interference from hacking, jamming and spoofing. Just as codes were intercepted and broken by early computers in Bletchley House under the supervision of Alan Turing, altering the path of the second World

War (and of information technology), so AI has significant vulnerabilities that could be exploited.

Geographical issues arise both in terms of machine learning and hardware production. Machine learning requires a lot of computation and energy, which means that it is unlikely to be able to be done on a mobile device. Thus, wireless connectivity is essential if it is to be used remotely. This brings with it many issues in terms of universal use and potential interference.

Another geographical aspect relates to the supply chain associated with semiconductors. In 2018 alone, semiconductor sales totalled over four hundred and seventy-six billion dollars. Semiconductor production is dominated by South East Asia, and China is rapidly increasing its production capacity. With China already dominating rare earth metal supply, it is beginning to dominate the supply chain for technology. Both the Obama and Trump administrations have blocked Asian companies from buying semiconductor manufacturing companies such as Xcerra, Qualcomm and Fairchild Semiconductor. The semiconductor supply chain has now become an issue of national security.

The issue of self-regulation is an interesting one in economics. As mentioned before, Adam Smith thought of this regulation coming from an invisible hand that emerges from society. This is meant to act to prevent selfishness and vice from overwhelming the process. Indeed, humans do show self-regulation. The question remains as to whether this is true altruism, for the sake of the tribe, or pseudo-altruism, ultimately in their own selfish interests, as neo-Darwinian thinking would have it. Indeed, the idea of selfishness being the fundamental drive, rather than some tempered, virtuous self-interest, was first put forward by Bernard de Mandeville, in 1705, in his poem, *The Grumbling Hive*, when he wrote "*Thus every part is full of vice, Yet the whole mass a paradise*" (Mandeville, 1705, lines 155-156).

So how do humans self-regulate? Other than Smith's concept of the invisible hand, more recent work has focused on feedback loops. Feedback loops must be continuous in order to track environmental changes and emergent events, as we would expect in systems theory (see Section V.8). Of course, the system can fail, and often has throughout human history. Complexity arises from there being numerous feedback loops simultaneously, leading to a need to trade-off between demands.

Whether the goals we ultimately pursue are innate and genetic, or stem from an emergent resonance within society is debated – a nature or nurture issue. If AI is to be truly intelligent, then this is an important issue. How would AI work through conflicting feedback and arrive at an outcome that was human-like? Without knowing how we arrive at our actions in relation to goal achievement, then how are we to design AI to do the same?

No matter how we approach the issue of goal orientation and motivation, we must reflect more fundamentally that there is a significant caveat. Modern humans have failed completely to embrace environmental feedback in their self-regulation. The evidence for this failure is all around, and becoming increasingly clear. And so, if we are to design AI to self-regulate, then we need a much more integrated model, one that uses environmental and societal feedback. This will be essential if decisions made by AI are to contribute to a sustainable future. This is a fundamental message of this book.

VII.2. AI and society

The complexities of society and the impacts of AI upon it are manifold. Here are some of the questions we need to consider. What do we mean by a good society? This is important in terms of any consideration of AI negatively or positively impacting upon society. Is there one global society, or many different types? If there are many different types of society, then AI could impact differently upon each type. Is society an emergent entity, or is it constructed from its building blocks in a reductionist fashion? This latter question matters as, if it is emergent, then any impact upon it, such as AI, will have unpredictable outcomes by definition. The study of social phenomena, sociology, has many different schools of thought, and within each school, different predictions are made for the impact of AI. Let us consider a few of these schools.

Symbolic interactionism holds that humans interpret the observed reality through a lens specific to their cultural and personal histories, replacing any objective reality with subjective symbolism. In other words, symbols, constructed by a person, replace reality. It tends to reduce society to individual interpretation as the basis for interaction and is a form of existentialism.

The symbolic AI approach that dominated AI in the Sixties and Seventies, forms a happy bedfellow for this school of sociology. Symbolic process models use the 'if...then' method, building from input to output. This reductionist approach reflects an emphasis in programming focused on the representation of an individual human actor, be they a chess player or a quiz show contestant. Furthermore, the individual human has been treated as the unit of currency, rather than some societal collective.

This approach, focusing on the individual as the creator of symbolism and, therefore, the building block, has been re-enforced by evolutionary biologists, who voraciously attack any suggestion of selection at the level of society (see for example, George Williams (1966) who, in his book *Adaptation and Natural Selection*, wrote *"group-related adaptations do not, in fact, exist"*). Selection is seen to operate at the individual or genetic (selfish gene) level, not at the societal level, be it a society of squirrels or of humans. Society is built from genes upwards, a strongly reductionist argument.

Socio-biology, based on this concept, has become a significant area of study, though concerns relating to eugenics and the selective removal of negative genetic material from a society through sterilization or execution, has made this subject extremely contentious.

Other reductionist approaches include *rational choice theory*, where behaviour of an individual is based on calculations made on the outcomes of that behaviour for the individual, and *game theory*, which utilizes similar cost-benefit approaches at the level of the individual actor. *Social exchange theory* shares similar foundations, utilizing assessments of potential rewards and punishments as determining our behaviour.

The reductionist schools of sociology stand in opposition to Adam Smith's concept of the invisible hand, an emergent principle, wherein we experience empathy or resonance with a societal characteristic, and then construct our own ethical framework from this resonance, guided by the invisible hand of societal influence.

Conflict theory, as construed by Karl Marx, saw power imbalance and authoritarian domination as the invisible fist, forcibly shaping societal order. This is a much darker theory, and resonates with fears that AI could become a tool of the powerful to control us, akin to some form of Orwellian big brother.

Functionalist theory, developed by Emile Durkheim, emphasizes that each element of society functions for the stability of society as a whole. Akin to the *Gaian hypothesis* in biology, which emphasises that all organisms and their inorganic environments contribute to a self-regulating system that perpetuates life on the planet, functionalist theory embraces systems theory, rather than reductionism. Here, output (societal stability) produces appropriate inputs. This has much in common with the more recently dominant school of AI, *neural network theory*.

Neural networks have come to dominate in AI, emphasizing connectivity. I compare this to the ball and stick models of chemistry. I remember in school building increasingly complicated models from balls and sticks. The balls represent the individual atoms in the molecule, whereas the sticks represent the energetic bonding between the atoms, reliant upon the interactions between the atoms that stemmed from the different characteristics of the atoms, and the electrochemical conversations between them. Different personalities exist across the hundred-odd types of ball (called elements) and this determines the structure of the molecules.

However more complex molecules like proteins can be further affected by the energetic contexts in which they were found, and some can change shape. For example, on the outer membranes of our cells, we have crucial structures called transport proteins. These can change shape and act like little machines, if supplied with energy. From a functionalist perspective, the sticks ultimately determine the spatial and functional reality of the molecule, not the balls *per se*.

This is quite different from symbolic modelling, which represents a ball-centric approach, based on symbolism and entities. Symbolism dictates where the balls should be placed, to definite rules, whereas the stick-based non-symbolic approach would focus on the interactions and emergent properties, wherein changing context can be accommodated and ambiguity is not as much of a challenge. In terms of societies, again we can either focus on the emergent characteristics resulting from the myriad of interactions (a stick-focused, non-symbolic approach), or a reductionist process of viewing the individuals building society and explaining its properties (a ball-focused, symbolic approach).

As we can see, there is a diverse array of theories for interpreting what society is, how it functions and, in turn, how AI might impact on it. AI has also been used to aid the study of society. *Artificial Social Intelligence* (ASI) is the application of machine intelligence techniques to social phenomena. The field really began at the Social Action and Artificial Intelligence Conference in Surrey, England, in 1985.

One thing is sure. AI has the potential to have huge positive impacts upon society. Let's examine some of these impacts.

Perhaps no aspect of sociology has greater potential for improvement through AI than public health. As more data relating to our health is computerized, from image scans to medical records and genetic analysis, medical big data is now a burgeoning business and the basis of medical research.

Fuzzy logic deals with degrees of truth, rather than true or false approaches. It offers advanced diagnostic potential, where complex interactions can produce a multitude of potential causes of a given set of symptoms. Individuals can often become overwhelmed when independently using search engines to match symptoms with putative causes. However, with advances in computing, such as fuzzy logic, these variables can be managed more capably than with the 'if/then' approach of symbolic programming.

It is unlikely that AI will replace a medical consultation completely in the near future, but, rather, will work with the medical practitioner, another example of intelligence augmentation. Taking initial medical histories and combing data from different aspects of a patient's life, such as work, social service interaction and counselling, is already being done by machine. Imaging and interpretation of scans is another area to which AI is ideally suited.

In terms of diagnosis, three areas have utilized AI more than others: cancer, cardiovascular and nervous system malfunction and disease. Other recent applications include restoration of control of movement in quadriplegia (Bouton et al., 2016), where the researchers applied machine learning algorithms to decode the neuronal activity of a patient and translate this into activating control of the participant's forearm muscles. Another application is the use of AI in retinopathy, where a smartphone-based device can photograph the retina, and then these images can be analyzed by automated analysis software using machine learning (Rajalakshmi et al., 2018). Such approaches allow rural communities in impoverished areas to have access to important screening tests at a greatly reduced cost.

In cancer diagnosis, an IBM Watson system, called Watson for Oncology, is a pioneer in this field (Lee, 2014), utilizing machine learning and using analysis of big data (specifically from the Memorial Sloan-Kettering Cancer

Center). While errors have been reported, this must be set in context. Around one and a half million human errors in medication prescription occur every year (Lee, 2014), while the average U.S. patient is expected to be harmed by diagnostic errors at some point in their life (Sakr, 2016). AI has the potential to deliver much more consistently correct prescription and diagnostic outcomes than humans.

Mining big data also has the potential to help in disease prediction. *Deep patient disease prediction* (from the alternative term for machine learning, *deep learning*), led to improvements in *precision medicine*, allowing patients to be monitored for potential health issues across a wide range of conditions (Miotto et al., 2016).

Other developments are already operational. *The Human Diagnosis Project*, or *Human Dx*, combines the world's collective medical intelligence with machine learning, allowing the global medical community to build a series of maps to aid diagnosis (Abbasi, 2018). Benefits include access across the world, including in some of the poorest countries.

The *RightEyeGeoPref Autism Test* by RightEye LLC, helps identify autism in very young children (12-40 months) with an eighty-six percent success rate (Southern Medical Association, 2017). Identifying autism early allows for early intervention, widely accepted as being extremely important. The test presents the child with two screens, showing two different programs at the same time, one with faces, and one with geometric shapes. Children on the autistic spectrum tend to spend more time looking at the program with geometric shapes.

Epidemiology is the study of the patterns and spread of disease through the human population. It is critically important and its relevance has been realized in recent avian influenza and Ebola virus outbreaks. Modelling and predicting the spread of disease allows us to respond. By monitoring social media, there is the potential to mine big data and model disease spread. Issues related to data ownership arise here as do issues relating to data quality. But certainly, epidemiology is a field where AI could play a very significant role (Wong et al., 2019). Similar applications are being investigated in environmental epidemiology, tracing the spread and impact of pollution (VoPham et al., 2018). The main barriers are regulatory (in terms of assessment of AI reliability/suitability) and freedom of information related issues, preventing the sharing of data.

Perhaps the most commonly used computer technology is social media. It allows us access to the wealth of knowledge available online, connecting the internet of things and providing social interactions across the globe for little cost. It is hard to believe that none of this existed thirty years ago. This connectivity can have downsides, but the upsides are also noteworthy. Connecting people together has, perhaps, the greatest implications for society. The statistics for Facebook are awe-inspiring. There are four million 'likes' every minute. Two hundred and fifty billion photos have been uploaded. One hundred million hours of video are watched every day.

The average number of connections, or friends, on Facebook is three hundred and thirty-eight, more than twice as high as Dunbar's number, one hundred and fifty. Robin Dunbar a British anthropologist, developed his number based on a relationship between primate brain size and social circle size. He defined it, in a very British way, as "*the number of people you would not feel embarrassed about joining uninvited for a drink if you happened to bump into them in a bar*" (Dunbar, 1998). I am not sure how many non-human primates unexpectedly join friends at a bar and order a pint, although I did have a rather unfortunate social interaction with a macaque, who grabbed my beer on a hotel veranda in the Surat Thani region of Thailand, drank it and then climbed into the roof timbers above my head and urinated on me. I don't know if he included me in his social circle, but he certainly partook of an uninvited drink.

The vast amount of data now circling on social media platforms supplies a potential goldmine of information if properly mined. This is what machine learning is so good at, not merely processing data, but learning from it and imbuing it with meaning. Thus, our ongoing interactions with our friends will undoubtedly contribute to a machine, somewhere, learning more about how we work, and what it is to be human, or human-like.

Social media is already being used during humanitarian crises. Artificial Intelligence for Disaster Response (AIDR) is a free and open source software that collects and analysis social media interactions, such as images and tweets, in a disaster zone, and can also send out messages. A walk-through is provided by Patrick Meier (Meier, 2014).

Transport is likely to be transformed by AI, as autonomous vehicles move onto our streets and start to impact on our ability to go places. Imagine someone who is blind or paralysed, now being able to travel so much more easily. Lorries could keep driving without the concerns over a sleepy trucker. The car could be sent on errands without a human on board. AI is already being used in UBER taxis. Smart cities are growing, where technology can take charge of traffic control.

Society can also be protected by AI, wherein identity theft can be more accurately spotted and stopped, banking fraud can be prevented and plagiarism, hacking and spam can be halted. Of course, the very processes of hacking, spam generation, data and identity theft and banking fraud could also be carried out by AI, but more of that later.

In terms of employment, working conditions could well change dramatically. AI promises to get rid of the four D's of the employment landscape: dull, difficult, dangerous and dirty jobs. Industrial and nuclear plants could be operated by robots. We have already discussed the fears of unemployment, but there are clearly advantages to having AI in terms of safer working conditions. A robot can be sent where it is dangerous to send a human (think mining, heavy industry and deep ocean operations. The job with the highest proportion of fatalities in the world is an underwater welder).

Criminal Justice and security are other areas where AI is already playing a role. AI is good at mining data, a key approach both in crime prevention and in detective work. AI is also used in predictions and modelling for distribution of policing efforts. Concerns exist around private power and public accountability (Who owns test data? Are police and criminal justice programmes using similar algorithms? Can a defense team access these algorithms?), and around potential bias within algorithms, training data and systems design (e.g. bias against ethnic minorities, females, the poor etc.) leading to a pernicious feedback loop (O'Neil, 2016).

Big nudging combines big data with nudging, a process of seeding our search results with particular suggestions or nudges. This has been used in elections and other areas of societal life. Cambridge Analytica harvested vast amounts of personal details of millions of Facebook users and then used it for political nudging in the Mexican presidential elections, the Brexit vote and the US presidential elections. Data manipulation also impacts negatively on collective intelligence, relying on the resonance and 'sympathy' of individuals for each other, within societal structures, such as the invisible hand of Adam Smith.

A significant concern arises with nudging. In order to appear invisible, nudges take the form of *self-resonance*, wherein your searches end up feeding you information that confirms your overall views. This can lead to social polarization, wherein groups no longer receive a balanced diet of input, but merely obtain data that reinforces their standpoint. This can lead to fragmentation and, ultimately, the breakdown of a functioning society.

As Helbing et al. (2017) put it, "*for collective intelligence to work, information searches and decision-making by individuals must occur independently. If our judgments and decisions are predetermined by algorithms...this truly leads to a brainwashing of the people.*" Collective intelligence requires social diversity. Helbing et al. (2017) observed: "*Pluralism and participation are not...to be seen as concessions to citizens but as functional prerequisites for thriving societies.*"

Forrest Mathews, in *The Ecology of Democracy: Finding Ways to Have a Stronger Hand in Shaping Our Future*, building on years of the Kettering Foundation's research, calls relational environments "*the wetlands*" of democracy, drying up with the spread of technocratic systems (Mathews, 2014). The challenge is to maintain societal empowerment while positively interacting with advanced technology.

With big data now rapidly expanding, there is pressure for a "right-to-a-copy" charter. The Swiss government is currently examining this idea, giving ownership back to the individual. The idea is to sanction digital self-determination, allowing individuals themselves to profit from the global data economy. In much the same way as an author is paid every time someone borrows one of their books from a library, or a songwriter is paid every time one of their songs is aired on a radio station, so each of us should benefit when our data is used by a company. Thus, rather than businesses selling our data to each other, we would become active actors in such transactions.

In society, *'fake news'* is a term that we are more familiar with each day. Yet AI has the potential to generate fake news through algorithms hidden deep within its programmes. If society begins to doubt their digital reality, or if certain actors convince us that something could be fake when it is not, it will become increasingly difficult to protect a fundamental value, trust. Concerns over the possibility that recent significant referendums and national elections have somehow been tampered with have created significant assaults on our confidence in democratic processes and in technology.

Such are the changes envisaged for society, that there is a call for a redefinition of humanity itself. Given the increased connectivity of humans as globally networked, changes in social interactions such as gaming and flash mobbing, the speed of communication in real time, knowledge access and issues relating to big data and privacy, the very fabric of society is transitioning to a new age. This is not necessarily negative, but it is different.

Another potentially tricky area relates to companion robots. Could we fall in love with a robot, and become romantically involved? How would this impact on our perceptions of what a relationship is, and could it impact on our relationships with humans? In 1975, the movie, *Stepford Wives*, based on the novel of the same name, written in 1972 by Ira Levin, explored this concept, where wives were turned into obedient cyborgs for their husbands' pleasure. With the children also converted to robots in the sequel, *Stepford's Children*, the town boasted of having the lowest crime and divorce rates in America: a cyborg-human paradise.

Can we attach emotionally to a robot? I've certainly cried a silent tear or two at the loss of a particularly loved car or building, inanimate though they were. But is this because I loved the inanimate object or because I was really mourning the loss of a physical link to memories related to the car or house? People attach to their pets. A companion robot would be different, generating its own memories. Think Singer in Star Trek, who dreaded being rebooted as he would lose his memories, not just his files. And attachment is a two-way street.

As foster carers for many years, my wife and I were continuously told never to form attachments to the kids we cared for. Yet it was impossible not to, even subconsciously, and we would cry our souls out as each child left for a new start in a forever family, adoption. It is human to attach. We just got better at dealing with it over the years. But when I started helping on training sessions with new foster carers, I made it clear that healthy attachment is a natural thing, not something to feel guilty about, or a failure in some way. I felt that the earlier training we had received was actually really counter-productive and wrong. So, we should expect to attach to our AI companions. But is this ultimately problematic?

Lucida and Nardi (2018) warn that companion robots could lead to a form of hallucinatory reality for the human, with the potential to result in social malfunction. Sparrow and Sparrow (2006) suggest that companion robots would be unethical due to the artificiality and deception that such friendship would entail, writing: "*What most of us want out of life is to be loved and cared for, and to have friends and companions, not just to believe that we are loved and cared for, and to believe that we have friends and companions, when in fact these beliefs are false*".

Even more intimate than the robot-human interface is the concept of Transhumanism. The World Transhumanist Association (WTA) was founded in 1998 and their website states that "*Transhumanists advocate the moral right for those who so wish to use technology to extend their mental and physical (including reproductive) capacities and to improve their control over their own lives*". If humans become technologically augmented, how will this change society? While more science fiction than science fact at present, it is by no means beyond the bounds of possibility that humans could become increasingly fused with technology, producing super-humans. If only the elite could afford this, the outcome could be concerning, with huge inequalities emerging.

VII.3. AI and the environment

As environmental damage continues to increasingly impact on the Earth system, the need to address this issue has become a priority in the twenty first century. Key to such efforts are the gathering and analysis of huge amounts of data, satellite image analysis dealing with system-level complexities, planning and monitoring. The message of this book is that these urgent tasks are perfectly suited to an AI-based approach, playing to all of its strengths. As we have stressed already, however, the solution space cannot merely be focused on single issue resolution, but must be prepared for trade-offs across each and all of the environmental, economic and societal arenas.

The World Charter for Nature by the United Nations (UN, 1982) points out that when "*potential adverse effects are not fully understood, the activities should not proceed*". This concept has been formalized in the *Precautionary Principle*, which states that "*When an activity raises threats of harm to human health or the environment, precautionary measures should be taken even if some cause-and-effect relationships are not fully established scientifically*" (Hayes, 2005). The precautionary principle is enshrined in EU law (CEC, 2000).

Of course, this creates difficulties. Given that all environmental problems are systemic and therefore emergent, cause-and-effect rarely applies completely. Destinations are unknown and adventitious. Newtonian science, or reductionism, is thus at a loss, and this is without including the economic and social complexities that also must be incorporated. Thus, we are more within the realms of *New Physics*, with its relativity and uncertainty principles.

Furthermore, is the uncertainty merely a matter of inadequate data? In other words, if we knew everything about everything, then would uncertainty, and with it, emergence, disappear? Whatever you think on that, the reality is that we do not know everything. And this lack of knowledge has trapped us before.

Think of the *'miracle'* of CFCs, developed as refrigerants (freon) by Thomas Midgeley Jnr. Midgeley Jnr also invented leaded gasoline. The combination of these two inventions led the environmental historian, John McNeill (2001), to refer to him as having *"had more impact on the atmosphere than any other single organism in Earth's history"*. The point here is that incomplete knowledge led to disastrous consequences in terms of skin cancer from the ozone destruction through CFCs, and the brain damage, particularly in children, resulting from leaded petrol. Jessica Reyes (2007), from the National Bureau of Economic Research, reported that following the Clean Air Act in 1970 and subsequent legislation banning lead-based paints and leaded gasoline in 1990, decreases in childhood lead exposure have been responsible for a fifty-six percent drop in violent crime in the 1990s.

That's where the precautionary principle comes in, recognizing the uncertainty and taking the better-safe-than-sorry approach. There are countless other examples of disastrous over-confidence in poorly understood processes, such as the drug thalidomide, or the deliberate import and release of exotic species that proliferate and smother out native species. Cases of the latter include the cane toad from South America, released into Queensland, the Burmese python, released from Asia into Florida, the Nile perch, released from the river Nile into Lake Victoria, the grey squirrel, released into Britain from America and Japanese arrowroot, released in the USA from Asia. All of these species led to the extinction or decline of many native species and the disruption of entire ecosystems, and, often, billions of dollars in economic damage.

However, the precautionary principle is not without its critics. H.W. Lewis (1990) claims that it amounts to a *"delusion of conservatism"*, wherein, by seeking to reduce negative risks to public health, it can actually lead to increased risks. He uses the example of strengthening aeroplane wings to make the plane safer, but this leads to a heavier, less manoeuvrable plane. He also refers to the situation of astronauts, trained for all conceivable incidents, meaning that they are less trained for the most likely ones because their time is spread across multiple scenarios. Frank Cross (1996) writes that *"Applied fully and logically, the precautionary principle would cannibalize itself and potentially obliterate all environmental regulation."*

Trade-offs will certainly prove challenging within precautionary principle thinking. How do you weigh up whether to use wind, nuclear, solar or coal power generation? Risks exist for all of them, in terms of pollutants, public health issues and environmental considerations. How would the precautionary principle solve this? Equally challenging is how to programme the precautionary principle into AI and at what level of strength.

However, the precautionary principle is meant to be just that: a principle. Thus, it should act as a form of moral canon, allowing us to carry out our scientific assessments of risk and then informing our decision-making. It allocates a responsibility and burden of proof. So, the company producing the product has the financial costs associated with proving its safety.

As Andrew Stirling (2007) puts it, the precautionary principle "*provides a general normative guide to the effect that policy-making under uncertainty, ambiguity and ignorance should give the benefit of the doubt to the protection of human health and the environment, rather than to competing organizational or economic interests.*" Uncertainty is part of any system. Ambiguity is a consequence of social diversity while ignorance is an outcome of a lack of complete knowledge. These are challenging issues for any AI involved in environmental problems.

Let's go back to the 4th October, 1957, the birthday of a huge technological contribution to our vision of a sustainable future. We find ourselves just outside Baikonur in Southern Khazakstan, but currently rented by the Russian Federation. Baikonur is a small city, nestled against the banks of the once mighty Syr Darya river, which was thought to flow from paradise. The river is now a shadow of its former self, its waters taken for irrigation of the vast cotton fields and polluted with uranium from the mines of Uzbekistan and Kyrgyzstan. A clue to the importance of Baikonur lies in the name the Russians use for it: *Zvezdograd*or *Star City*.

It was just before 7-30 pm local time, when a rocket blasted into the skies, marking the launch of the first artificial satellite, Sputnik 1, into space. It would orbit the Earth for around three months, and marked the beginning of space exploration by humans. Since then, some eight thousand four hundred satellites have been launched, of varying complexity, some of which have gone beyond our solar system. However, most of these satellites orbit the Earth. Currently the United Nations Office for Outer Space Affairs estimates that as of the beginning of 2019, there are 4987 satellites orbiting the planet (UNOOSA, 2019).

These satellites have many uses, including communication, navigation, television broadcasting, weather forecasting and environmental monitoring. And it is the latter that is so important for our vision of an ecologically-intelligent AI for the future. Measuring environmental change at a planetary level from the ground is very difficult due to problems relating to infrastructure, temporal coverage, travel, political instability, spatial coverage and labour costs. For example, measuring if an ice sheet is shrinking in a remote part of Antarctica would require an extremely challenging expedition to be mounted.

Satellites can measure all sorts of things across the globe from space. Ocean temperatures, sea level rise, and ocean and wind currents can all be measured. Pollution in the air and in water, forest fires and global assessments of ecosystem functioning can be assessed. Biochemical, physiological and biophysical properties of ecosystems, as well as biodiversity, structure and composition of the species present be monitored.

In terms of sustainability, the four key indicators of progress can be assessed: natural capital, ecosystem functioning, biodiversity indicators and habitat intactness. Satellites can use advanced acoustic remote sensing to measure biological and geological characteristics of the sea bed. Measurement of atmospheric dust allows indirect analysis of soil erosion.

In fact, satellite imaging can offer a full, real-time health check of our planet. There are barriers out there, including reluctance of governments to share data, and a lack of communication between ecological and remote sensing communities.

Definitions of what actually constitute a particular habitat change can create issues. Take, for example, forest ecosystems. In the 1990s, the United Nations Food and Agriculture Organization (UNFAO) defined a forest as consisting of a minimum of ten percent bamboo and tree coverage, associated with wild fauna. In 2005, this was changed. Wild fauna and bamboo were dropped from the definition, and now trees had to have a minimum height of five metres. This would rule out elfin forest for example, a stunted ecosystem that grows at high altitudes, and many bamboo-dominated habitats. This means that it is difficult to compare forest health over time, as the parameters of what defines a forest have changed. The day before the new definition came into being, you might have been a forest. The next day, you were no longer a forest.

The rather clumsily titled *Group on Earth Observations Biodiversity Observation Network*, though with a catchy acronym (GEO BON), have a stated vision of *"A global biodiversity observation network that contributes to effective management policies for the world's biodiversity and ecosystem services"* (GEO BON, 2019), and brings together the remote sensing and ecology communities to advance this. The use of the term 'ecosystem services' is unfortunate, though widespread, as we have already seen. The group have defined ten variables that should be focused upon in terms of satellite telemetry, and the approach is aimed at meeting the *Aichi Biodiversity Targets*, the core outcomes of the *Convention on Biological Diversity*, a multinational agreement signed in Rio in 1992. These variables are: species occurrence; plant traits (such as specific leaf area and leaf nitrogen content); ecosystem distribution; fragmentation and heterogeneity; land cover; vegetation height; fire occurrence; vegetation phenology (variability); primary productivity and leaf area index; inundation.

Much work still remains to be done. Relating the broad-scale remotely-sensed data to fine-scale ecological events and processes can be difficult. Research on American amphibian populations has shown that remotely sensed data may fail to detect changes in conditions that threaten animal populations on the ground or may detect these changes too late to alert managers and rangers (Gallant et al., 2018). Thus, a multi-platform, multi-scale, integrated approach is essential to avoid these issues.

Another issue relates to forest assessments. At present, oil palms and industrial forests (those planted by forestry companies for commercial reasons) cannot clearly be delineated from natural forest recovery. Thus, the displacement of native forests by oil palms may not be detected by satellite data (Muller-Karger et al., 2018). However, as new satellite technology with greater capabilities are introduced, these issues will be addressed. Higher spatial, temporal and spectral resolution, along with higher radiometric quality (together called H4 technologies) are becoming available. Satellite data is also perfect for AI to analyze and compare and contrast.

One issue with ecosystem functioning is that it doesn't easily lend itself to quantitative economic comparison. In other words, it is difficult to put a monetary value on it and to evaluate whether or not things are improving or degrading. How do you evaluate the sense of identity or a feeling of wonder as you gaze at a landscape? What about average global temperature change? How much was the last dodo worth? Nature changes relatively slowly, and can have tipping points, as we've mentioned before, where there can be little or no measurable change and then suddenly there can be a large swing. Because Nature is a complex system and emergent in character, we cannot extrapolate current trends into the future with any confidence. So even if we can measure it, we cannot predict anything from this.

Without quantitative analysis, it is difficult to assess and monitor change. Not everyone can benefit equally for cultural benefits from the environment, as cultures differ as does access. If you are one of the million people living in the Dharavi slum in Mumbai or one of the seven hundred thousand people living in the Kibera slum in Nairobi, you may not have the opportunity to benefit from recreational or educational rambles in the countryside. So how can we put a value on Nature, when there is such a disparity and inequality in the benefits on the ground?

And should we be putting an economic value on the Earth system anyhow? It can certainly be argued that equating natural capital to economic capital makes no sense, since they are not equivalent. Anything you measure is merely part of the whole system, and so its value cannot be teased apart from the value of the entire living world. Since this cannot be bought, sold or replicated, then there is no point in attempting to value it, as it is invaluable.

However, one benefit of at least putting some kind of figure on a particular habitat such as a wetland, is really to remind us how priceless it is. As long as we realize this, then the valuation can play a useful role. Insurers will place values on paintings such as the Mona Lisa (insured for eight hundred and thirty million dollars in 2018), but if it was ever destroyed, it would be irreplaceable, and all the money in the world could not restore it.

Nature is unique, a product of three billion years of evolution, and can self-heal. It is synergistic and self-regulating. It produced us and the millions of other species that have lived or continue to live their lives on this planet. It is clearly unique in the solar system, and, possibly in the universe. Nature is beyond value, and cannot be partitioned off into aliquots of cash.

ARIES (Artificial Intelligence for Ecosystem Services) is a network-based modelling platform, mapping service flows, benefits and beneficiaries, in order to visualize, value, and manage the ecosystems on which the human economy and well-being rely upon. Aside from the issue of the use of the word 'service', it has a potentially useful place in the re-integration of humans into the Earth system. It aims to inform policy makers of the importance of ecosystem functioning while clearly communicating potentially complex issues in an understandable way for non-experts.

The ARIES approach uses machine learning and pattern recognition, along with fuzzy logic, to incorporate uncertainty, in order to trace the flow of benefits from Nature to the human end-user. ARIES uses four stages: sources, uses, sinks and flows. Firstly, in the 'sources' stage, it assesses possible ecosystems sources for a particular benefit. Next, in the 'uses' stage, it assesses who could benefit from the particular service and what is the need. The 'sinks' stage involves examining what impacts increasing or decreasing source provision has on a sink. For example, in a flood prevention situation, water supply might be reduced or water use increased. Finally, the 'flows' stage examines actual flows through various paths that connect the source with the sink.

It was Gaston and O'Neill (2004) who first posited the question in a paper of the same title, "*Automated species identification: why not*?" Using AI to identify species has been a focus for a number of years. Species identification usually requires specimens to be examined by expert taxonomists, usually based in natural history museums. However, with AI, combining image recognition software with machine learning, the potential to identify species even from a photograph or two, is becoming possible.

Image processing has also been used to identify butterfly species with ninety-eight percent certainty (Kaya et al., 2014), a notoriously difficult taxonomic challenge. Powerful taxonomic tools such as the Pl@ntNet app now combine the power of social networks with automated taxonomy tools. Pl@ntnet has been used by four and a half million users since its launch in 2013. Yet only one percent of submitted images have actually been validated, meaning there is huge potential for future work.

Such apps are a central plank in the *Citizen Science movement*, where everyone can contribute to research projects through data acquisition. With an estimated four billion mobile devices worldwide, the use of apps in Citizen Science opens up an exciting new world. This represents a potentially giant investigative network, with four billion sets of eyes and ears, observing the world and feeding these observations back into some form of electronic medium.

This really does have the potential of connecting AI with the living world in a comprehensive way. *The Great Koala Count*, *iBat*, *Instant Wild* and *Zooniverse* are all examples of these apps. The latter, initially focused on astronomy, now has nearly two million registered users, all involved in a wide range of Citizen Science projects, from pelican ecology, through manatee call research to earthquake monitoring.

AI is also in use in pest and disease control programmes, where it can accurately and rapidly identify which of the many potential pests is causing a particular disease outbreak, through image processing (Fedor et al., 2016). This is essential for proper modelling and control.

A significant challenge in terms of monitoring progress in sustainability activity is the inherent stochasticity within Nature, where droughts, disease and weather events can massively impact ecosystems and the ability to monitor them (Legg and Nagy, 2006). I remember one of my PhD students, whose research relied on continuous fieldwork, having her whole thesis jeopardized by the foot-and-mouth disease outbreak in the UK in 2001, resulting in her not being able to access her field sites for a year due to the closure of public rights of way. Since most research contracts are only three to five years in length, a significant natural disruption can completely ruin sustainability monitoring programmes.

Remote sensing of habitats using electronic sensors on the ground, on drones or on satellites, and wildlife telemetry of individual organisms by tagging, allow huge amounts of data to be collected without the need for humans to be there, and this data is the perfect food for AI to number-crunch its way through. Human expert analysts are and will continue to be essential for this process, but the AI can massively reduce the amount of wildlife data they need to examine. This is a nice example of a complimentary working relationship, or technological augmentation. Deep learning is already being used to inform ecological decision-making, such as in identifying suitable habitats for potential re-introductions of rare species (Hirzel et al., 2011), as questionable as this might be if not integrated within a fully-functioning food web. For example, you should not re-introduce a herbivore without re-introducing its predator.

AI has also been suggested as a means of addressing one of the great paradoxes in sustainability: the need to reduce human influence on Nature versus the increasing need for human management of wild places. Bradley Cantrell and colleagues (Cantrell et al., 2017) suggest a form of alternative Turing Test (which originally checked to see if a human could discern differences in a conversation with another human and with an AI robot), where humans enter an autonomously created and managed 'wild' space and think it is natural. While far from achievable at present, the vision is to have evolving algorithms that learn how natural processes lead to ecosystem development and resilience, and then assist Nature in this process while countering negative human impacts.

Of course, one of the great problems with conservation biology is that it attempts to freeze things at a particular evolutionary point, and the choice of which point to recreate is very subjective. I call this the *Garden-of-Eden complex*, where, as Joni Mitchell sang *"we've got to get ourselves back to the garden"* in her 1970 classic, *Woodstock*. But what is the garden? Was it ten thousand years ago, before we settled and began the agrarian revolution, turning our backs on the hunter-gatherer lifestyle? Was it forty thousand years ago, in the middle of the last ice age, when woolly mammoths gave us a run for our money? Was it sixty-five million years ago, just before the dinosaurs and flying reptiles were wiped out? Was it three hundred and seventy-five million years ago, during the Devonian, the Age of the Fish, when our aquatic friends diversified to rule the world?

I also refer to this compulsive, yet baseless desire to recreate some habitat from the past, by re-wilding, as *Golden Age environmentalism*. This is a very strange concept, given that you cannot go back in time. Conservation efforts, whether carried out by man or machine, will never really produce natural systems, since these are emergent, not constructible, given that Nature is a system. Surely, we should, instead, allow Nature the freedom to diversify as it sees fit, focusing on withdrawing our impact upon it. AI could play a very important role here, rather than replacing us as the great controllers of all.

Data-driven modelling (DDM) is the process of identifying useful patterns in data. As data sets become huge (literally, big data), there are multiple opportunities to analyze and learn from patterns within this data. However, the challenge is to separate the wheat from the chaff. DDM approaches allow for the production of inter-disciplinary models, incorporating social, economic and environmental data sets. These in turn provide the opportunity to explore more holistic solution spaces. The issue of trade-offs is still missing, however, but it certainly provides extremely useful information in terms of policy development and decision-making.

Invasive species pose a significant threat to biodiversity, and it is not only their identification, but a prediction relating to where they could thrive upon invasion, that is important. As the climate continues to destabilize and change, areas previously impossible for an invasive species to spread into, because the temperature and rainfall regimes were incompatible, now become suitable. Fuzzy cognitive maps explore the interactions between various aspects of this issue, and elucidate the strength of impact of each aspect on the others.

Demertzis et al. (2018) used these maps to predict changes in rainfall and temperature in Greece as a result of climate destabilization. They then explored when a number of potentially invasive species, currently prevented from succeeding in Greece due to hostile environmental conditions, would be able to invade, and where. This powerful approach allows counter-measures to be taken earlier, which, in terms of invasive species, is critical in order to prevent devastating impacts on ecosystem function and agriculture.

VII.4. Technology and sustainability: Bellicose bedfellows or Romeo and Juliet?

It is clear that AI offers many positive elements to the three arenas. However, each of these has been considered in isolation, but as we have seen, there are repercussion for each arena if we try to optimize any one arena. People have in the past shown how AI can help in various tasks, like collating diversity data. But what is needed is a much more elemental transformation. Because our planet is a functioning system, we must expect trade-offs and sub-optimality.

Given that technology is fundamentally targeted at achieving outputs desired by humans, particularly in terms of production efficiency, optimization and profit, it has been questioned whether technology can ever really deliver a sustainable solution. This has divided opinion, with two extremely entrenched camps defending their positions. Given that this book sets out a vision of how AI can provide a path out of environmental and social collapse, it is now time to address this issue head on.

Should we use technology to achieve sustainability or would we be selling our souls to the devil? As we have already noted, advocates for a sustainable future tend to be split down the middle when it comes to technology and its role in our futures. On one side of the debate lie the technophiles, who believe that technology is the sabre of progress, allowing us to build the New Jerusalem, and pave the road to a humanist utopia, transferring our grand visions into a physical reality. This is a powerful Enlightenment view, wherein our future lies in our own hands and we can design sustainability through technology.

This view is also called *weak sustainability*, defined as development that meets the needs of the present generation without compromising the ability of future generations to meet their own needs. The focus is on needs, and the three arenas are interchangeable, meaning that it doesn't matter if the environmental arena diminishes, as long as other arenas can replace it. Technology is seen as key.

Technology, it is argued, can actually replace Nature in terms of providing ecosystem services for our future survival, whether it be through genetic engineering, cloud seeding, iron enrichment of the oceans (to draw down CO_2) or biomimicry, where we borrow ideas from Nature and implant them within technology. *Terraforming*, the transformation of another planet to make it Earth-like and thus inhabitable for humans, takes sustainability to another level, where we can create new worlds.

The word 'terraforming' was first used by Jack Williamson, using the pen name Will Stewart, in a short story he wrote in 1942, called *Collision Orbit* (Stewart, 1942), but the concept found its origins a decade earlier, in a book written in 1930 by Olaf Stapledon, entitled *Last and First Men: A Story of the Near and Far Future* (Stapledon, 1930). Stapledon suggested that Venus could be terraformed. The first non-fiction work referencing terraforming as a serious pursuit was by Karl Sagan, the astronomer, in 1961, again focusing on Venus as the target. Later, attention would shift to Mars.

Advocates of technology as the means of delivering sustainability emphasize that the relationship between technology and society is key to building a better world (see Johnson and Wetmore, 2009), and that the role of the environment can be replaced. One school, *instrumentalist technology*, insists that technology can be transferred to developing nations without any ill effect. This school believes that a combination of genetic engineering, solar energy and the internet will eradicate poverty in the world. Banuri et al. (2001) state that "*The development and diffusion of new technologies is perhaps the most robust and effective way to reduce GHG [greenhouse gas] emissions*".

Another school of thought, *colonization*, sets out the benefits of using technological interventions to modify and maintain an alternative natural state that makes it more useful to humans, a sort of terraforming of our own planet. Of course, the universal roll-out of technological solutions across the planet, even if it were neutral, as claimed by the instrumental technologists, would hasten the disappearance of diversity, with all of the threats and moral implications that this path brings. Economic globalization and the internet of things already pose such a threat, and, as we shall shortly discuss, one of the greatest challenges to AI is the place and significance of diversity, and how this can be protected within a connected world.

Christopher Potter sums up this approach in his book *How to Make a Human Being: A Body of Evidence*, when he claims, entirely incorrectly, that "*Humans never were part of Nature. We were always part of technology*" (Potter, 2014). Potter would appear to have forgotten that we spent by far the majority of our existence (around one hundred and ninety thousand years out of the two hundred thousand years of our time on the planet) embedded within Nature as huntergatherers, not technologists. Also, we emerged out of Nature, not technology. Fundamentally we are reliant on Nature, and the nutrient, water and energy cycles have always flowed through us, as they have through the rest of Nature. We are Nature. Also, many other species use so-called technology, but we would not consider that they were "*never part of Nature*".

Finally, the *substantive technology* school stresses that technology exists within its own bubble, and is not neutral. No person or institution can slow or halt it. Technology introduces new values, transforming social life and culture. Take transport as an example. When people only had horses, travel was very limited, with populations not being very mobile. People would be born, live, work and die within a small locality. Families took responsibility to care for their elderly, who also had a role as the village elders. Several generations would often live in the same house. Now, with people moving often to find employment, families are less nuclear, and the care for the elderly often falls into the capable hands of social services. Significant social change has therefore been delivered by changing transport technology, be it cars, buses, trains or planes. And technology is globalizing.

Furthermore, it is often underlined that we have, in ways, been using technology for a very long time, from tool-making and fire use, to books, binoculars and cameras. We have been inventing things to help us for as long as we have been around. However, other animals and plants have also been using tools, fire and other technology for much longer than we have, but not in an isolating way. Leaf-cutting ants have advanced agricultural technology, preparing a mulch from leaves and then growing a crop of fungi on the mulch. Genetic diversity of the fungal growth is maintained.

Eucalypts have highly flammable bark that peels off, helping to create fires to which their seeds are immune. The seeds of many plants only germinate if there is smoke in the air or if they are burned. This use of fire as a signal works well, as it means the seedlings only start growing when they know that their neighbours have all been incinerated. Seabirds use stones as anvils to break open shellfish, as do otters. Beavers construct lodges.

Coral animals (*Cnidaria*) capture small algae (dinoflagellates) inside their own cells to provide them with sugar. This incorporation of another life form as technology is carefully managed and monitored, so that the algae don't reproduce too much and take over. This is done by limiting the access to nutrients inside the coral animal's cells, so that the algae don't have enough nutrients to reproduce. An interesting aside is that when coral reefs are exposed to fertilizers coming from the land, nutrient levels rise and corals die because the algae multiply excessively within them, forming 'black coral' (Kelly et al., 2012).

In a weak sustainability paradigm, benefits do not need to be global in terms of economics. Provided the mean global wealth and welfare increases, those countries doing the best (i.e. the developed nations) can compensate the losers. Not only can each arena compensate for the others, paying our way out of trouble, but the inequalities, while maintained, can be ironed out too. It's a morally questionable position, but lies at the heart of this form of thinking.

On the other side of the argument lie the *strong sustainability* advocates, who firmly hold that Nature cannot be replaced by technology. Strong sustainability can be defined as a development that allows future generations to access the same amount of natural resources and the same environmental and social capital as the current generation. Immediately there is a problem here, in that, given the damage that already exists, this definition translates as the maintenance of the damage as well as the capital, not allowing for restoration. Perhaps a better definition would include a reference to restoration of social and environmental capital, with a reduction in economic capital, that currently has parasitized both the social and environmental arenas.

The reasoning underpinning strong sustainability arguments lies in the reality that Nature is an emergent system, and you cannot control emergence. The full spectrum of Nature colours far more than merely ecosystem services. Indeed, it was never there to service humans in the first place. It embraces essential contexts of our being.

Biomimicry, for example, would encourage us to take aspects of Nature and place them in a technological context of product design. This represents an anathema to strong sustainability advocates, who would claim that taking structures out of their context and embedding them in a completely different context is a strategy of doubtful value, if not downright irresponsible. They also point to the thesis of substantive technology, wherein technology is a runaway train, hurtling out of control and requiring a fundamental system change to derail it. Furthermore, it could be claimed that technology has contributed to wealth inequality, social inequality and the environmental crisis.

Within any discourse on sustainability, there is a nexus where intergenerational and intragenerational justice face off against each other. By this, I mean that the desires, rights and existence of the present generation may have implications for the next generation, and *vice versa*. This is a trade-off, not merely spatially, but across time. Of course, we cannot tell exactly what consequences our actions will have on the future configuration of the Earth system, because of the emergent nature of complex systems. We do know that damaging the environment and failing to build a sustainable society will have negative impacts. We also know our future generations will rely on us curbing our excesses in terms of resource use.

Do we then say that people in poorer countries cannot have the lifestyles that we have enjoyed for decades, because this would threaten intergenerational justice? Can we play the trade-off game with something as fundamental as social justice? Can we justify the terrible damage done to new-born babies around the mines that provide the rare earth metals that are used in wind turbines in order to make the planet a greener place (Li et al., 2013)?

Also, maintaining current 'standards' in terms of the lifestyles enjoyed in the 'developed' world may protect the future generations of the West and North, but may significantly jeopardize the future generations of the South. This is why intergenerational justice must be spatially as well as temporally homogeneous and consistent. And this will require a complete restructuring of the planet in terms of equality.

For we cannot have global social justice without a regional re-alignment. In other words, the Northern 'developed' nations will have to meet the Southern 'developing' nations half way (or possibly a bit further south) in order to achieve global intergenerational justice. And this is before we add to the mix the essential issue of environmental justice (i.e. the rights of the environment). Of course, ignoring environmental sovereignty will likely only lead to our own removal from the mix, thus simplifying things, as the three arenas become one, with no social or economic arenas left to worry about. This is what we can call *environmental singularity*.

It is clear that the current, dominant economic model represents the broader Enlightenment philosophy of individualism, progress, economic growth and development. However, it is also clear that this model is not delivering the promised social impacts, as demonstrated by increases in inequality in spite of increasing GDP, both globally and nationally. The treadmill of production theory is shown to be a more accurate measure of the relationship between the three arenas than is the ecological modernization theory, as we have already noted. Furthermore, and more fundamentally, the environment is being degraded by the current economic programme. This has significant and punitive implications for society.

In terms of transitioning to a sustainable future, there are really only two conclusions to draw from this:

1. We need to change the dominant philosophical framework that underpins our use of technology and our abuse of both the societal and environmental arenas;
2. We need to change the trajectory of the most dominant technologies embracing these philosophical changes, in order to deliver a sustainable future.

Let's consider each of these two points. The first point relates to the dominant philosophical framework, Enlightenment thinking within the humanist tradition. Human rights are very much focused on the individual in Western traditions, but as we have discovered, this is not so in other cultures, such as the *buenvivir* philosophy of the Andes, or the sub-Saharan African *Ubuntu* philosophy. Here, there is a focus on society as the unit of empowerment and sovereignty, tightly tied to the environment.

Furthermore, not every nation on Earth is at the same point in its structuring. Some are in the Information Age, others still in the Industrial Age, some are agrarian and a very small number are still hunter gatherers. Thus, there is a great heterogeneity across humanity, both philosophically and structurally. While environmental damage plays out at the global level in terms of ozone depletion, ocean acidification and climate destabilization, other forms of environmental collapse are much more local, such as pollution from mining, dead zones in the coastal seas and soil erosion and salinization.

However, we have also seen that Western economies export the ecological damage that ensues from their own resource-rich lifestyles to other nations, meaning that the cost of such lifestyles is not visible to the consumers based in the 'developed' nations, and that they do not suffer the sociological and environmental consequences of these lifestyles. Therefore, it is more difficult to recognize how damaging our lifestyles, and the philosophy underpinning these lifestyles, are.

Solutions that appear to be working fine have implications elsewhere, but what you don't see doesn't matter. If a tree falls in a distant forest, and we can't see or hear it fall, then has it really fallen at all? Supply chains disappear into the mist, and what's in the mist doesn't count. Shelves stacked with products appear to have been beamed down from some awesome alien planet. And things are cheap due to mass production, so it really can't be doing any damage. Cheap means disposable, and cheap means easily gotten. Cheap also means mass consumption, mass environmental damage and a lack of respect for the planet.

Our products have hidden origins and hidden consequences and lack the financial tags that would reflect on the true environmental costs. It's all too easy, and it is designed to be so. Our Enlightenment philosophy has no place for Nature other than as a source and sink. It also has no place for society, because individualism rules. It's a toxic mix, where economics delivers all, whether it be a trickle-down effect, or the promise that each of us can become wealthy one day.

The Enlightenment thinking of Condorcet and others proclaimed that Nature would not hold us back. We thought we were the train engine. Without the burden of the dirty, shoddy worthless wagons, filled with those aimless and visionless passengers with their teeth and claws, all of them lacking intelligence and sentience, we could steam ahead along the tracks of progress, unhindered and unfettered, and make our way to the bright and shining city of Utopia, where we would live forever and realize the humanist dream. But we made one mistake. We weren't the steam engine, we were just a very heavily ornamented, smug yet vulnerable carriage ourselves. Disconnected from the true engine, we are doomed to slide backwards and become derailed.

One of the greatest flaws in Enlightenment thinking lies at its very core. It is a reductionist framework. In other words, it ignores systems thinking. Reductionist thinking is a building block philosophy, where you can add or take away bits and alter the structure. It's a bit like a game of Minecraft©. Adding and taking away blocks is the Enlightenment's way to solve problems.

Recently in Scotland, beavers have been re-introduced, mostly illegally. Their populations are growing rapidly and the beavers are cutting down young trees that provide essential flood protection and habitat stabilization. Scottish tree populations were already under pressure from poorly managed rabbit and deer populations. The foxes who predate the rabbits were hunted by humans to near extinction, while the predators of deer are long extinct in Scotland.

Ironically, it was argued that the beavers would ease flooding and control water flow. However, unless you introduce a predator of the beaver (a bear, lynx or wolf), the beaver population will grow completely out of control, as it is doing, leading to unheralded destruction of our vulnerable forests. Lifting a beaver-shaped block and attaching it to the fragile building that is Scottish forest ecology is a disastrous move.

There are other issues too. The European beaver is actually made up of 8 sub-species. The decision to move one of these, the Norwegian sub-species, to Scotland is extremely dubious. At the end of the last ice age, the most likely sub species was the French one (*Castor fibergallicus*), because the land bridge was to France, not to Norway. Norway was covered in ice just like Britain, and so the Mediterranean refuges were much more likely to form the source of our beavers. DNA analysis of beaver pelts from British beavers could be done to check this, but, surprisingly, this hasn't been carried out.

A study by Walter Durka and colleagues (Durka et al., 2005) stressed that the geographically nearest form should be used, and that there is a huge danger related to re-wilding in terms of the future evolution of the beaver. In evolutionary terms, species start off as sub-species, and to intervene in this process by moving groups around the continent, we are potentially impacting on the future direction of beaver evolution.

The context of an organism is everything. Every herbivore needs food and a predator for a natural balance to be achieved. Without the predator, the population will, always, spiral out of control without culling. But what is the point of bringing the beaver here just to be culled? There is a moral issue here. Beavers are intelligent mammals, and so if our actions deliberately lead to us needing to cull, then this is not a morally acceptable approach. Culling is also not a simple process.

By killing particular beavers, we will not necessarily replicate the natural force of predation, because we may not kill the weaker beavers, but rather the stronger ones (for example those that disperse most). This can lead to a genetically weakened population, and thus to all sorts of genetic problems.

Suggestions that beavers that spread too far will be culled or returned home, in order to prevent their spread, are also extremely worrying. For example, after two years, juvenile beavers naturally migrate from their natal site. They can travel up to one hundred miles (one hundred and sixty kilometres). The reasons for this are to reduce population load at a particular habitat, and, even more importantly, to prevent inbreeding. If we do not allow these migrations to happen, inbreeding depression will occur, and this can lead to terrible deformities. Do we really want this? To avoid this, we would need to allow these migrations. If we do this, then there can be no control on the spread of these creatures, and they are likely to encounter roads, probably acting as a significant hazard to drivers and themselves.

One beaver family destroys three hundred young trees in a single winter. Tree regeneration is difficult enough with rabbits and deer already putting unacceptable pressure upon young saplings. However, the beaver removes bark containing the main transport system for sugar, thus killing the tree. Grey squirrels also do this, and so in combination, this is an unacceptable problem.

To use the beaver as a means of terraforming (changing a habitat into one that works for us) is an irresponsible and vulgar strategy. The Harlequin ladybird was introduced to act as a predator on aphids, and now its population has run out of control in the UK and threatens many of our native ladybirds. The cane toad, as we have already noted, was introduced into Australia to kill insect pests of sugar cane. But it didn't eat the pests, instead gorging on many endangered insects. Its poisonous skin secretions have killed many other animals. Biological control and biological engineering are never likely to work because the ecology is usually too complex to model and predict. It is like a car mechanic attempting brain surgery. In fact, it is like a brain surgeon attempting brain surgery – the outcome is not secure.

Yet to add insult to injury, the current Scottish government, pandering to the eco-activist voters, have now made the beaver a protected species. It's a delusional and, frankly, unforgiveable ecological crime that has the potential to destroy what is left of Scotland's forests, while increasing the flood risk, since the trees that remove water from the soil are being destroyed.

Conservation efforts, whether carried out by man or machine, will never really produce natural systems, since these are emergent, not constructible, given that Nature is a system. Surely, we should, instead, allow Nature the freedom to diversify as it sees fit, focusing on withdrawing our impact upon it. AI could play a very important role here, rather than replacing us as the great controllers of all.

The application of inappropriate metaphors is a very basic flaw in so much of our current thinking on sustainability. Take the circular economy, a leading concept gathering momentum since the Chinese government adopted it as part of their five-year plans and the EU also placed it at the heart of their efforts. Built around the fiction that Nature is efficient (eco-efficient), running a no-waste circular economy which is optimized towards productivity, it has been used as a metaphor for a new approach to economics.

However, there are a number of problems, as discussed in a recent paper by Skene (2018). Most fundamentally, Nature does not run a circular economy. In fact, Nature is extremely wasteful and inefficient, and for two reasons. This is because Nature is a system, which, if working properly, is sub-optimal at every level within it. Every day, the Sun rises and pours currency into Nature's economy, as sunlight is converted into chemical energy through photosynthesis. And every day, all of the chemical energy is used up, most of it being released as waste.

As ecosystems develop and become more complicated, more energy is needed to maintain them. If you need proof of the waste, look at a typical food pyramid in Nature. At the bottom there are lots of plants, making sugar. But as you go up the pyramid there are less and less organisms at each level. If Nature had a circular economy and was efficient and optimized, these would be food cubes, not pyramids. But they aren't cubes, because Nature is wasteful.

If you wanted to run the human economy on the lines of Nature, you would need a large fleet of alien spacecraft depositing megatons of gold onto the planet's surface every day. That is what the Sun does, day after day after day. If you stopped the supply of the golden sunshine, say by blocking the light with dust in the atmosphere, it would not be good. Ask the dinosaurs. Oh, hang on, you can't, because they all went extinct when the sunlight was blocked in just that way, sixty-five million years ago.

The natural world does not have a circular economy, and so this false metaphor should not form the basis of our plans for a sustainable future. You can't just lift blocks of stuff from one system and stick them in another, even if the blocks were real. In the case of the circular economy, the blocks didn't even exist in the first place. Let's not be like the king in the children's story, who paid a huge amount of money for a new suit and convinced himself that he had bought the most beautiful outfit in the world ever. The nakedness, danger and futility of the circular economy as a workable solution must be recognized. Efficiency is not the way to re-integrate with the Earth system. In fact, drives towards efficiency are what got us here in the first place.

And so, it is evident that Enlightenment thinking will not be able to deliver sustainability, neither will misplaced metaphors and a reliance on exchangeable capital. More fundamentally, we need to move away from a concept of sustainability that is anthropocentric. If we frame sustainability as producing human intergenerational justice, we miss the point.

Rather it is the continued functioning of the Earth system that must take centre stage, and then the elucidation of a path that allows us to continue as part of such a system. Thus, a functional, sustainable philosophy must focus on systems thinking, and what that means. Once we understand this, then we can find our place within it, and the place of technology in delivering this. We now move to the final sections of this book. This puts flesh on the bones, detailing what we need to change, how we can do this, the barriers that exist and how we overcome them.

SECTION VIII Imagining a New World

"Imagine no possessions
I wonder if you can
No need for greed or hunger
A brotherhood of man

Imagine all the people
Sharing all the world

You, you may say I'm a dreamer
But I'm not the only one
I hope someday you will join us
And the world will live as one."

–John Lennon, *Imagine*

Welcome to Section VIII. We've explored AI, discovered ecosystem intelligence, faced the crisis confronting us and surveyed the three arenas. It's now time to bring these strands together. How can AI save the planet? In this section we explore exactly how this can be done. First, we'll delve further into why human intelligence has led us to such a point in our history, before identifying what needs to change. We next examine two case studies: the honey economy of the Ogiek people from Kenya and the feather economy of the St Kildans from Scotland. There are rich lessons to be gleaned from both of these societies. We redefine the meaning of the Smith's invisible hand. Next, we will consider how not to save the planet, before detailing the central role for AI in this great project.

VIII.1. The swallow whose nest was stolen: A salutary tale

We have danced our way through history, in an orgy of destructive excessiveness that would make Bacchus himself cringe. The Greek god, famous for music, wine and ecstasy, represented freedom and hedonism. Mixed with our feelings of entitlement to do whatever we want to our planet for our own gains, lies power through wealth. This toxic mix has led to the devastation and corruption of the essential ecosystems that underpin our existence.

It was Aristotle, around 320 BC, who named the four elements, from which, he believed, all things were made: water, air, earth and fire. And, as we have seen, the hydrosphere, atmosphere, soil and energy flows have been cataclysmically altered and poisoned, wither by our insatiable mining of the natural resources of our planet, the release of waste materials or the overloading of crucial energy fluxes through the natural circuitry of our food webs that were not designed to cope with such levels of energy.

For clear evidence of this, we could turn to the Intergovernmental Science-Policy Platform on Biodiversity and Ecosystem Services (IPBES, 2019), that reports that seventy five percent of the land-based environment and about sixty six percent of the marine environment have been significantly altered by human actions. Of note also is the fact that these trends have been less severe or have been avoided in areas held or managed by indigenous peoples and local communities.

In terms of oceans, thirty three percent of marine fish stocks were being harvested at unsustainable levels, with just seven percent harvested at levels lower than what can be sustainably fished. Since 1980, the extraction of non-renewable and renewable resources has doubled. Plastic pollution has increased tenfold since 1980. Since 1700, some eighty-five percent of wetlands have been lost. These wetlands form vital protection for shores, store vast amounts of carbon and capture nutrients that would otherwise flow into the sea, leading to dead zones.

One million species are threatened with extinction in the coming decades. Some half-a-trillion dollars of crops are at risk because of a decline in pollinator species. There is a powerful irony here, in that a major contributor to pollinator decline, particularly related to bees, are the pesticides sprayed onto these very crops to improve productivity.

One member of the panel, Professor Sandra Díaz, summarized the situation as follows: "*Biodiversity and nature's contributions to people are our common heritage and humanity's most important life-supporting 'safety net'. But our safety net is stretched almost to breaking point*" (Shieber, 2019).

However, you don't need a United Nations report to comprehend the changes we are causing. There are small ecological disasters unfolding all around us. We live in a rural village in Angus, East Scotland. It's an idyllic place, with lots of wildlife. Ten years ago, the old derelict mill opposite our house was converted into a business. The results were devastating. There had been a pair of swallows, migrating from South Africa each year, traversing the Sahara, Morocco, Spain and France on their incredible journey. They nested in a carefully constructed mud nest, accessed through a small hole in the stone wall of the old building. Their arrival each year was a special event and inspired a feeling that all was still right with the world after all.

The new owners of the mill plastered up the innocuous looking hole in the wall. The next spring, a swallow was found, exhausted and near death, lying beside the building. Despite valiant efforts, the bird couldn't be revived. It had survived a journey of six thousand miles, flying across hostile environments. Our swallow didn't just settle for reaching the first barn on the white cliffs of Dover. It kept flying another eight hundred miles to reach us in rural Scotland. Indeed, although swallows only live for an average of four years, their offspring will often come back to the nesting site of birth, and so a lineage is established. Certainly, the nest in the old mill had been occupied since we had first moved there, some fifteen years earlier.

But in rural Scotland, in an old mill beside a river and a forest, humans had ultimately deprived it of its entrance hole in the wall, and, exhausted from its travels, it gave up. It just didn't have the strength to build another nest of straw, mud and saliva, nor to search for an unlikely abandoned nest. After all, the bird was probably born here a few years earlier and this was the only home it knew.

This swallow hadn't just been let down by our economic activities. Legislation in Scotland only protects swallow nests that are occupied. If builders want to get rid of a nest, they just need to wait until the end of the nesting season. And so, these amazing birds are being let down by our laws. Imagine if someone with a holiday home, that they visited for a few months each year, found that it was illegal for anyone to break into their property and destroy it, but only if they were actually living in it at the time of the robbery.

Back at the mill, things deteriorated. The business, which was now using noisy machinery, began to disturb not only the structural but the audial environment. Soon the red squirrels stopped coming to the bird table, where they had shared seeds and nuts with their feathered friends. In the forest beside the old mill, there were no longer the tell-tale signs of chewed pine cones. The squirrels had gone too. Over the next year, we noticed that there were no longer tree sparrows visiting the bird table, and then the number of dunnocks declined to zero.

Walking along our road before and after the changes in the mill, the differences would have been imperceptible. Yet because we live here, we saw the changes first hand. This one business had impacted massively on the local wildlife. But when we think of the impact of a rapidly expanding and resource-hungry human population globally, the IPBES report shouldn't surprise us. Yet it does, because travelling around the countryside, everything seems in order. A snap shot reveals birds singing, rabbits hopping and trees growing, much as they always have done. But in reality, most of the landscape you see is artificial: commercial forests, gardens filled with exotic species, industrialized agriculture and buildings no longer suitable for a weary sparrow that has lost its home.

The loss of a few species here and there may not seem cataclysmic, yet the overall impact is huge. There is so much evidence pointing to the current trajectory. This is a steeply spiralling decline as we continue to wear down our natural world. Yet with the planet's human population still on the rise, and with the average human using increasingly larger amounts of natural resources, Nature, sick in form and function, is being pressured like never before.

And this is why all of the recent international scientific reports are pointing to a current crisis. We can no longer deny that our plane is going to land. There was only ever going to be this outcome. Resources are running out. Air, water and soil pollution are reaping a deadly harvest among humans and other life forms. Soil is eroding away. The coastal seas are overfished and cannot support recovery of fish populations. Unless we act now, urgently, immediately and within a longer-term plan, we will crash land. What could this look like?

Without pollinators there would be no more fruit. Without sufficient soil, or with salinized soil, there will be a collapse in agriculture. With increasingly pressurized agricultural systems, intensification will only exacerbate the issues. Increased use of fertilizers will lead to poisoned drinking water and dead zones where fish cannot survive. Climate destabilization will lead to increasingly violent storms, sea-level rise with concomitant coastal flooding, and droughts that decimate food productivity and increase soil erosion.

Sea level rise will displace millions of people from areas such as Bangladesh and the Pacific Islands. Meanwhile, the disappearance of Himalayan glaciers will lead to rivers such as the Indus, Sutlej, Ganges, and Brahmaputra all but disappearing, impacting one billion people or more. When water and food begin to become scarce, civil unrest and forced migration are to be expected, destabilizing entire regions.

VIII.2. Blinded by the bling: Dashboard dogs and a disappearing sea

How do we change direction? Can we turn left at the next roundabout? Can AI provide the equivalent of a GPS system for the human journey? Problematically, there is no roundabout and the signs for alternative routes to avoid the apocalypse are overgrown and hidden by piles of human comfort, ease, desire and pleasure. We can hardly see through the windscreen of our three-litre diesel roadster because of all the bling we've hung on the mirror and the assorted toy dogs with nodding heads and rhinestone cowboy boots on the dashboard.

Urgent indicators outside of the speeding vehicle are just not noticed. We're heading down progress highway, underpinned by Enlightenment philosophy and encouraged by economists, commerce and politicians alike to embrace the route ahead. Within this techni-coloured bubble with its nodding plastic dogs seemingly commanding us to put the pedal to the metal, everything seems just fine.

It's a small track, just off to the left. In fact, there have been countless opportunities to turn off the huge sweeping ribbon of highway that cuts swathes across the planet, innumerable reminders that we need to change things. Yet we have continued, consolidated and concentrated our efforts into shaping the world for our own advantage, transforming it into a plaything, reducing it to a state of complete enslavement and abuse.

We are driven by selfish, individualist desire, encouraged by an economic model unfit for use. We sit high in our luxury penthouse at the top of the tree of life, but the tree is dying and soon the trunk on which the penthouse rests will give way. It's already creaking, but we can't hear it because of the new sound system we've just installed. We're at a much bigger crossroads now. No small track this. The signs can't be ignored. We've got to brake and change direction. So how do we do that?

We've seen clearly how Enlightenment thinking, and its subsequent implementation in the economic arena, has caused massive damage to the Earth system, threatening our future generations with insurmountable difficulties. While Adam Smith was one of the leading philosophers and architects of this theory, we have also seen that his work was misunderstood. As we shall now consider, within Smith's model lies perhaps the salvation of our species, and the redemption of our place within our ecology. And enveloped within that, lies a spectacular role for artificial intelligence. Let me explain how AI can save the planet.

We have noted that the invisible hand emerged out of society and reached into economics. But there is a third arena, one that was ignored by Smith: the environment. When we put these three together with an invisible hand now spread across all three, we have the potential to establish a truly new age, an age of re-integration.

Economics and society impact each other. Lord Lionel Robbins defined economics as "*the science which studies human behaviour as a relationship between ends and scarce means which have alternative uses*" (Robbins, 1935). Thus, human behaviour lies at the heart of this definition. Indeed, Robbins suggests that economics defines human behaviour "itself as the relationship of" means and ends. While we have seen that Smith sets out society as providing the calming voice to economics, economics also impacts strongly upon society. The flow of resources through society is a core element of economics.

Employment is determined by economic activity. Money, stemming from economics, sadly still has significant impacts on health and education outcomes, in that wealth inequality is closely tied to health and education inequality. This is particularly true when the society-economy nexus is missing the invisible hand of Smith. An economy unchecked is damaging to societal function. Also, a healthy, wealthy and educated society would, certainly in Smith's view, only be good for the economy, both as a market and as a workforce. So, there is feedback between both arenas.

Economics and the environment also have a powerful relationship. If we go back to Lord Robbins' definition of economics, the scarce means refers mostly to environmental resources, be they metal ores, oil, forests, land or fresh water. The economy is dependent upon the raw resources available.

Some countries have greater natural resources than others. In our history, the Fertile Crescent became the birthplace of modern civilization. The fertility came from good soils and, most importantly, a bounteous supply of fresh water from the three great rivers, the Tigris, the Nile and the Euphrates. As we have mentioned earlier, the Agrarian Revolution, where humans settled, built cities and developed an economy, relied on the grain harvests to underpin these other developments.

No longer did humans have to hunt and gather, though these activities undoubtedly still played a role until the development of widespread animal husbandry. The settled ones could rely on a generally consistent supply of food from the land, with excess grain being stored. Agriculture, since that time, has been central, both to the expansion of human populations across the globe and to the damage that now threatens us. This should not surprise us, because, ultimately, agriculture is all about energy production, in the form of carbohydrates, upon which we rely. It has become our own little food web and has allowed us to escape the natural world but we have lost our ecological intelligence in the process.

Yet the environment has a huge effect on economics too. This impact comes in two forms. Firstly, increased scarcity of resources, as supply peaks and then declines, represents a significant threat to us and our economics. Almost fifty years ago, Nordhaus and Tobin (1972) pointed out that *"the prevailing standard model of [economic] growth assumes that there are no limits on the feasibility of expanding the supplies of non-human agents of production"*. What they are referring to are the natural resources that underpin the supply chain of industry. Yet we know that this assumption is wrong. These resources are extremely limited.

Even so called 'renewable' resources are threatened because of habitat erosion. As we have already noted, fifty percent of topsoil has been lost in the last one hundred and fifty years. In China, soil is being lost fifty-four times faster than it is being formed, with eighteen kilograms of soil lost for every kilogram of food produced (Skene and Murray, 2017). Over half the world's wetlands have been destroyed, due to draining for building or agriculture.

In 2010, the European Commission (2010) reported that fourteen essential raw materials were now in critically low supply. By 2014, this list had lengthened to twenty materials, with the addition of chromium, borates, phosphorus rock, coking coal, magnesite and silicon. Others included graphite (essential for electric vehicles), platinum (essential for electric vehicles), Uranium 235 (nuclear power stations) and the rare earth metals (essential for computer technology, cell phones, rechargeable batteries, military applications and wind turbines) (European Commission, 2014).

What is noteworthy is that most of these resources are central to a low-carbon economy, where we attempt to move away from fossil fuels, not only for obvious environmental reasons, but because oil itself is running out. Phosphorus supplies are also running short, with peak phosphorus thought to occur in 2030 (Cordell et al., 2009). This will devastate agricultural economics, as phosphorus is essential for healthy crop growth. While oil can be replaced by other energy sources (at least, if shortages of key raw materials are ignored), there is no substitute for phosphorus, and it cannot be made by humans.

An interesting way to understand the impact of our overuse of materials is the concept of the Earth Overshoot Day (EOD). This is the date in a given year when we have consumed all that can be resupplied by the planet in that year. The earlier the date occurs, the earlier we overshoot the planetary limits. In 1970, EOD was 29th December, meaning that the planet was more or less able to replenish all that we had taken that year. By 1988, the EOD was 15th October. By 1998, it was 30th September. By 2018, it was 1st August. Another way of looking at this is to consider that in 2018, we would have required 1.7 Earths in order to have sustainable consumption.

A final point is important. Natural resources are always harvested in an identical fashion. The cheapest, most easily acquired supply is taken first, in what is called high-grading. When this runs out, more challenging sources are mined, with increasing energy, environmental and social costs. For example, with oil, a move to deep water drilling and suggestions of Arctic and Antarctic explorations have begun, with much greater pollution risks.

Gold mining of lower quality ore requires more fresh water to extract and produces much higher levels of cyanide, a key side-product of the extraction process. Thus, as resources deplete, environmental and social damage increases. Intensification of agricultural practice has led to soil erosion and compaction, reducing yields. This in turn requires greater intensification, causing more damage. Thus, the overshoot of consumption over replenishment will continue to accelerate unless significant change occurs.

The second impact of a diminishing environment on economics relates to the financial costs of ameliorating the damage caused by our activities. Soil erosion in the USA is costing over forty billion dollars per year in lost productivity, sedimentation and eutrophication (Halopka, 2017). Damage to wetlands globally is estimated to cost between six and twenty-four billion dollars per year. It is estimated that the cost of coral reef loss due to sea temperature rise will be one trillion dollars (Hughes et al., 2017).

Climate destabilization as a whole has recently been calculated to cost the economies of the world over twenty trillion dollars by the end of the century (Burke et al., 2018). A recent collapse of bee populations across the world, thought to be mostly because of pesticides used in agriculture, has been estimated to have cost one hundred and ninety billion dollars, according the United Nations Environment Programme (UNEP) (Wilson, 2013). The introduction of non-indigenous agricultural pests due to human activity is estimated to cost one hundred billion dollars per year.

Changes in the environment have fundamentally shaped the history of humankind. From the K/T mass extinction that led to the rise of the primates, to the Eocene that saw our major crop plants dominate, even our earliest origins have been shaped by environmental change. The Akkadian civilization of Mesopotamia became so reliant on irrigation that changes in climate led to their collapse. The inhabitants of Rapa Nui (Easter Island) deforested their little island in the middle of the Pacific Ocean, leading to societal collapse. Such collapse is often marked also by a cultural shift. In Rapa Nui, this was a shift from ancestral worship, symbolized by the Maui statues, to Nature worship, signified by the birdman cult. This shift reflected the realization that Nature, not the ancestors, was ultimately the determiner of their fate (Skene, 2011).

Humans have undoubtedly changed the environment too. The Aral Sea, formerly the fourth largest lake in the world, has disappeared in just sixty years because of irrigation schemes for farming. This has changed the entire climate of the region. Further back in time, the advent of fire in the aboriginal populations of Australia is now thought to have led to a major change in the vegetation and the megafaunal extinction around fifty thousand years ago. Human impact can be local (e.g. the deforestation of an island) or global (the recent rise in carbon dioxide levels, beginning perhaps three hundred years ago, and sea level rise).

And so, the environment, the global economy and human society all interact with each other. Furthermore, human, environmental and economic systems are all emergent and part of a greater emergent whole, the Earth system. Yet there is a significant issue here. While humans rely upon economics and Nature, and while economics relies upon humans and Nature, Nature doesn't need either.

Often the economics-society-Nature nexus is viewed as three overlapping circles, as if, somehow, they were all equally significant and dependent on each other. But this is blatantly misrepresentative. The arenas of human society and economics sit, both physically and metaphysically, on the foundations of the environment and are completely reliant on it for their existence, whereas if both ceased to exist, the environment would still be standing, as it did for the three billion years before humans were on the planet.

We may create a sixth mass extinction, as many feel we are doing at present, but Nature will survive, as it has so many times before. For we are *Homo vulnerabilis*, dependent on the environment for our existence. If things aren't just right, we die. Indeed, much of the technology we have developed has allowed us to increase our population far beyond the natural carrying capacity of our planet. We need heaters in winter, air conditioning in summer, supermarkets, pharmacists, malaria shots, antibiotics, hugely damaging industrial and agricultural output, all to help us spread across the planet.

Change, it is agreed, is essential and urgent if we are to avoid an existentialist crisis. Fundamental to this is the functioning and interactions of our economy, society and environment. The cohesive integration of these three arenas is an essential pre-requisite for human sustainability.

And this is where AI and the invisible hand of Adam Smith come together to offer a way ahead. As we have said before, a system can only properly operate when there is integration, real-time feedback and sub-optimality. It is this sympathy, as Adam Smith put it, this invisible hand, virtuous self-interest but operating across all three arenas, that will deliver solutions.

The invisible hand in this book is a far greater vision than that of Smith. For it straddles all three arenas, bringing them into resonance. Nature already does this. It is a brilliantly smooth-running machine, with resilience and sympathy coursing through its veins. We don't need to change Nature. But we do need to plumb back into this great circulation system, graft ourselves on and embrace it.

So, what is this hand that can do so much? This book proposes that it is artificial intelligence, based on ecosystem intelligence. Artificial intelligence can be the communicant, the arbitrator and the connective portal facilitating the environment-society-economy nexus.

VIII.3. What needs changed and what change do we need?

There are two aspects of change that we need to consider. Firstly, what needs to be changed in order for us to continue enjoying our planet? Secondly, what needs to change about us in order for us to make these changes?

In Section VI.2 we explored the main threats to our continued existence. They relate to soil erosion, salinization, eutrophication, diversity collapse and climate destabilization. Of course, these things are outcomes of a few basic issues: how we grow our food and how we manage our supply chains. If we approach this under a traditional, empiricist, reductionist umbrella, we could identify a series of problems, list them and attempt to tick them off as we solve them. Because we are dealing with a system, the tick box approach is seriously flawed.

Nothing is damaging the planet more at present than our food production, through the industrialization and technological intensification of agriculture, silviculture and aquaculture (fisheries). Humans need to eat a lot because we are multicellular, large and warm-blooded. However, the intensification of production has been partnered with increasing waste. Some forty percent of food produced is thrown away between farm and plate. If we didn't waste this food, we would need much less damaging production methods.

The production methods that hurt the planet most are intensive fertilizer application (leading to eutrophication and carbon dioxide release as a result of the energy needed to make nitrogen fertilizer), intensive irrigation (leading to salinization), soil compaction and degradation (leading to soil erosion), excessively large numbers of herbivores (leading to soil erosion and methane production from their digestive tracts), excessive use of pesticides on genetically modified crops (leading to the death of insect pollinators), the draining of wetlands and deforestation in order to gain more farm land.

Each of these methods is hugely damaging to the planet. Combine this with overfishing, the huge environmental cost of transporting food around the world and the use of antibiotics in farm animals and we are heading for an agricultural apocalypse that we need to avert. Agricultural practice must change, and change immediately. The problem is that industrialised and technological agriculture is a huge money maker, with some of the world's largest companies involved. These companies have huge political influence, as do the farming communities that fund them. However, this doesn't matter, because urgent change is essential.

Fundamentally, we need to farm better, eat less damaging produce (both in terms of the health of the planet and of ourselves) and eat more locally produced, seasonal food. We don't need to eat tropical fruit in temperate regions, nor do we need to eat summer vegetables in winter. In terms of farming, we need to return to the fallow year, where fields are planted with nitrogen fixing plants like clover, and then at the end of the year, this clover is ploughed back into the soil. This provides fertilizer and organic matter, stabilizing the soil, capturing carbon and reducing the need to apply artificial fertilizers. Increasing amounts of organic matter also increases the water-holding capacity of the soil, helping to buffer against flooding and drought.

Agroforestry should also be used. Here, rows of trees intersperse the crops, meaning that the trees help hold the soil together, preventing erosion, while controlling water levels and providing habitats for birds, which can help control insect pests without destroying pollinators (as pesticides do). Planting borders around fields also provides useful wildlife refuges, while preventing any excess nutrients leaving the fields. Here, plants like nettles and lupins can capture nitrogen and phosphorus respectively. Each year these borders can be cut and ploughed back into the field, again providing slowly released fertilizer and organic matter. This is all simple stuff, that used to be done regularly. It restores the soil, while displacing harmful methods currently used.

Although fields are non-productive for one year in three, provided we reduce the huge amount of food we throw out, then this will not be a problem. Furthermore, by reducing the amount of meat that we eat, we need far less land for grazing, and therefore we can grow more crops anyway. Because of food pyramids, we lose ninety percent of the energy available in the form of crops if we use the land for herbivore production (this is because by losing ninety percent of the energy from grass to herbivore, we require ninety percent more land to harvest the equivalent energy. Thus, it makes a lot more sense to eat more vegetables and fruit, and less meat. Eating less meat is also a healthier lifestyle choice for us.

AI can play a huge role in the greening of agriculture. Changes will be sweeping. We are talking about another agricultural revolution, but one that needs to happen. Remote sensing, wherein we can gather data on the nutrient levels escaping into fresh water and the oceans, monitor the health of our soils and track soil erosion, can provide the feedback needed to ensure that changes we make are having the right outcomes. Precision application of fertilizer, wherein remote sensing can detect which areas of a field need more or less fertilizers, can hugely reduce fertilizer run-off, by only applying fertilizer exactly where it is needed (Abit et al., 2018).

Being able to measure crucial indicators of ecosystem health will inform us of how well we are doing. Having vast networks of sensors all connected through the internet-of-things, integrated and feeding into an AI system charged with disseminating and analysing this data, will provide a powerful management system. We need such a system to guide us in our planning and implementation at the scale required. AI will be an essential player in this and many other aspects of this age of re-integration, but more on this shortly.

VIII.4. The chains that bind: Taking responsibility for our footprints

Supply chains are central to our agricultural and our non-agricultural resource use. For any given product, materials are often sourced from across the world. There are a number of issues here. Firstly, the materials in a product we purchase in our local store will have come from somewhere else. That means that the pollution generated from the production process will also have impacted elsewhere. Children may have been involved in manufacturing the product, and the workers may well be underpaid or living in virtual slavery. Mining of metals used in so many of our high technology products occurs in poor countries where health and safety is a low priority, both for the workers and for the communities living around the mine. Short cuts are taken. There is also a huge carbon footprint associated with the transport of the supply chain across the world.

And so, as we wander down to our local shopping mall, oblivious of the cries of hunger, impoverishment and abuse associated with the shiny, sparkly, brightly packaged items in the window displays, we fail to recognize that we are accountable for a good share of that suffering. Supply chains bring with them responsibility. The chain links us with every human, every village, every forest and every river that it runs through. Yet we are electively ignorant.

Of course, we could request a full ecological footprint of the product and its materials, the social cost and the environmental legacy of everything associated with our new purchase. But we don't do this. We just hand over our credit card and pop the product into the trunk of our car.

This has got to change. Unless we accept responsibility and ownership, rather than exporting the toxic, polluting price of our lifestyles to other countries and other people's children and landscapes, such denial will prevent us from making the changes urgently needed. Of course, the degradation of these far-off lands, socially and environmentally, will come home to roost eventually. The damage we are creating also impacts on our food chains, as many of our food products come from these same countries. Air and water pollution spread across the globe. So even if you are fairly xenophobic and selfish, this still concerns you.

To sort out the supply chains, we need information. What came from where? What impact did the mining and production process have socially and environmentally? Are we happy to share responsibility for this?

Imagine each day a large sign was carried around behind you everywhere you went. The sign carrier follows you into your business meetings, your child's school, the round of golf and your social event in the evening. On the sign is a list of all of the negative consequences of your lifestyle. A child died of an asthma attack due to the diesel pollution your car generated. A woman was burned to death making clothes that you bought, in a factory in Bangladesh whose fire doors had been blocked by boxes of materials, stacked high. A landslide wipes out a small village because the forest had been cut down leading to soil erosion and flooding, the wood of which was used to make your new dining room suite. Those rare earth metals in your mobile phone? The pollution around the mining sites in China led to significant liver disease among the villagers.

The sign now shows a slide show, with pictures of the devastation, photographs of the people impacted and a short bio of each. A child who was crushed while operating a loom in Southern India (used to make a nice woven rug you just bought) had been the best cricketer in his school and had dreams of a great future, lifting his family out of poverty from the income of a successful sporting career. There's a photo of his sister, distraught, at his funeral. You lower your head. *"I just didn't realize…"*.

We need information. With the internet-of-things and AI, we can easily access information about supply chains, but we need manufacturers to face their responsibilities too, examine their supply chains, and make this information freely available. The designers behind the products also must embrace much greater accountability. It is their designs, after all, that dictate what type of materials are utilized.

And so an informed production process can be easily generated if we want this to happen, where everyone knows the true costs of the products we buy, from the material supply chain, the human and environmental footprints, the ecological and social cost of their use during their lifetime (such as the diesel fumes from the car) and the ecological and social cost of recycling or dumping the object at the end of its life. This is essential if we are to become responsible consumers.

Thus, AI has, once more, a central role in the transformation of our supply chains, both in terms of informing us of the costs of such chains, and in monitoring the transformation of these chains into environmentally and socially positive processes. Imagine if each product had an empowerment score, expressing how the supply chain had helped build resilience in the natural and human communities involved.

Fundamentally we need consistent, informative monitoring and feedback, allowing us to understand how our actions are impacting the Earth system. Since the system into which we need to re-engage is emergent, AI can monitor the complex outcomes, allowing us to work with the uncertainty and surprise that are part and parcel of emergence. Systems such as the Earth system have so many interacting components that it is impossible to know exactly how they will respond. This is why trying to work out exactly what will happen in the future is impossible.

The planet isn't an extension of the genetic code, as Richard Dawkins (1982) suggested in his book, *The Extended Phenotype*. It isn't a model made out of building blocks, that can be added to or subtracted from. Rather, the Earth system is an outcome of myriad interactions, direct and indirect, with everything impacting on everything else. Like non-symbolic AI, it is a black box, where data enters and we can't be exactly sure what will result.

To manage emergence is not to control it but to prepare ourselves for surprise. The closest ally of emergence is feedback. If you want to cope with an ever-changing, dynamic and surprising world, the best you can do is to know what is actually happening. Real-time continuous feedback allows us to fully grasp the implications of our actions.

As a student, at the end of the academic year I remember being given a feedback form, where we were asked to highlight negative and positive elements of our learning experience. By then it was too late. Modules were completed, exams sat and nothing could be done about the year. Sure, it would help the next cohort of students, but more fundamentally it ticked the box of 'good teaching practice', the box that asked if the students had been involved in the design and running of the course, and if teaching and learning were active two-way streets.

Without constant, live-streaming monitoring, it is impossible to know what effects we are having. And AI, in combination with the internet of things, provides the perfect channel for such feedback. Never in the history of humanity has such a set of tools been available, and never have they been more needed.

Reliable real-time feedback has always been difficult to obtain. The water is muddied by self-interest, ulterior motives and tactical filtering, even if there is a willingness to listen and to learn. The other problem is that our hearing has been dulled. Why listen to Nature? Surely it has little to say that is of relevance. And it is this attitude, born out of the ideological principle that only humans are the wise ones, the top of the tree, the best in class, and so only we have the solutions needed. Only humans have the abilities and intelligence to plan and organize the rescue of the planet, and only we can save ourselves and everything else from death.

Yet the reality is very different. Everything will not die. The Earth system has always been its own physician. Of course, species that have run their course or that do not remain plugged into the system do go extinct. It happens all the time and has happened throughout the three billion years of life on Earth. But the planet has survived much greater devastation in times past.

Some seven hundred and fifty million years ago, there was a huge glaciation event, where ice covered most of the land, even at the equator. It would last almost two hundred million years and came close to extinguishing life on Earth entirely. A massive volcanic event eventually ended it, pouring vast amounts of carbon dioxide into the air and warming the planet. Yet the Earth system recovered and shortly after the great melt, complex animals emerged as can be seen in the Burgess Shale in the Rockies.

Fast forward to two hundred and fifty million years ago, and once again, life was almost extinguished on our planet. Almost all of the trees on the planet died, *en masse*, and ninety eight percent of life was wiped out. This was the mother and father of all extinctions, the Permian. Prior to the extinction, fossil layers are filled with spores, plant and animal remains. After it, all that were found were strands of fossilized fungi. This *fungal spike*, as it is known, represents death at an awesome scale, as the fungi feasted on the mortal remains of the lost.

This is an apocalyptic vision, one that could unfold again, but the point is that life went on, even after the blackboard was almost completely wiped clean with the duster of mass extinction. About one hundred million years later, dinosaurs were wandering around, birds and mammals were taking their first tentative steps and a myriad of plants thrived on the surface.

And the key thing to note here is that this recovery was not managed by a UN charter, an intergovernmental panel nor a bunch of academics, generously funded by a philanthropist billionaire. No, Nature sorted things out itself. The system got to work and a new, vibrant, living planet was created. Even if humans had existed two hundred and fifty million years ago and had survived the Permian extinction, we could not have done this. And it is the same today. Restoration, recovery and recuperation are much more safely handled by Nature itself. We can help, but only by giving Nature the space and time to recover.

However, we suffer from the curse of our ancestor: *Homo habilis*, the handy man. The handy man was a member of the same genus as us and appeared around two million years ago. Wherever remains of the handy man were found, so were a scattering of tools. Either they were constantly carrying and using tools, always died possibly from a tool-related incident (a likely endgame for myself) or were buried with their tools. Whatever the case, this was the first major human-like tool user. A doer, a maker, a fixer-upper.

DIY man was born. And we've been doing it ever since. This became a need, a condition, wherein our first response to anything is to act. I'm always reminded of *The Wacky Races*, a cartoon that was very popular when I was a child. In it, eleven eccentric characters competed in a series of races. One, *Dick Dastardly*, and his dog, *Muttley*, were nasty bits of work, always trying to cheat. Time and time again, their dastardly plans would fail, and Muttley would start to laugh at his owner. And it was at this point that Dick Dastardly always uttered the words, often while trapped under the wreckage of his latest ruse, *"Don't just stand there, do something"*.

It's the blight of humans to always feel that we need to pro-actively intervene: *Homo habilis*. Yet the opposite is usually true. Nature has managed for most of its time on Earth without such interventions, plans, codicils and legislation. It is a highly sophisticated system with complex communication and functioning. We should do less, withdraw our pressure and interference and let Nature get on with it.

St Kilda is a tiny group of islands, inhabited for the last five thousand years by humans. We'll examine it in detail shorty, for it has much to tell us. Home to many rare species, this rugged landscape is the remote outpost of Scotland, some forty miles beyond the Outer Hebrides and one hundred and ten miles west of the Scottish mainland. And it is a rare sub-species on the island that is of interest presently, the St Kilda wren.

The humans all left the island in 1930, and a team of ecologists were sent from the University of Cambridge, led by John Buchan and Tom Harrisson, to see what should be done in terms of the conservation of this tiny, indigenous bird, that had been closely intertwined with the islanders. Their report ended with the conclusion that "*Natural enemies are absent, with the one exception of man. The species will be best preserved by being left entirely alone*" (Harrisson and Buchan, 1934). These wise words apply to the entire Earth system. We are, generally, the only problem.

But what does it mean to leave things alone? It means not trying to manipulate Nature. Rather we need to work within the Earth system, fighting alongside it in the trenches, reducing our negative impact and becoming part of the system. This is where AI can play a transformative role. If we envisage re-connecting with Nature, then we must have open communication channels. Satellite and drone remote sensing allows us to monitor the system's health. We can assess ecosystem structure in terms of vegetation height, fragmentation, distribution and heterogeneity. We can measure characteristics of species populations such as disease prevalence, occurrence and demography. By monitoring atmospheric dust, we can evaluate soil erosion. We can monitor crucial aspects of ecosystem function such as fire occurrence, productivity and inundation.

Land-based listening devices, that measure temperature, salinity, current, winds, water chemistry and soil properties, allow us to examine detailed environmental processes. All of this provides feedback, and AI will be essential in terms of handling such a vast array of data. Then, as we gradually reduce our negative impacts, we can gauge how the Earth system is responding. Based on this, we can work towards the goal of restoring sovereignty to Nature, allowing it to self-heal.

Healthy ecosystems are diverse ecosystems, providing the functional redundancy that produces resilience. Also, natural ecosystems operate best with low levels of nutrients, limiting how much energy can pass through the system. This is essential. One of the most devastating things we have done is to increase the amount of nutrients, such as fertilizers, that are circulating today, mostly thanks to agriculture. This kills ecosystems because of excessive energy flow. So, we need to curb release of nutrients. A concerted effort is needed to wire all of this technology together.

This vast array of feedback will also act to re-engage us with the Earth system, in terms of being able to visualize what is happening on our planet. The technology already available, or soon to be available, is awesome. Satellites, drones, the internet of things and AI combine to provide the perfect basis for our salvation, but it takes one more thing: a transformation of our worldview. Because these same technologies, if combined with the currently flawed and disastrous worldview, will lead us more quickly to our demise, accelerating the downward spiral to the exit pipe at the bottom of the bathtub. And in our history, we have always used new developments in technology to support our damaging philosophy of humans first and only humans. Hence, it is down to us. We need to make some decisions now. We are at the crossroads. What are we going to do?

The Israeli historian, Yuval Harari, puts it like this: "*Sapiens regime on earth has so far produced little that we can be proud of… [D]id we decrease the amount of suffering in the world? Time and again, massive increases in human power did not necessarily improve the well-being of individual Sapiens, and usually caused immense misery to other animals… Moreover, despite the astonishing things humans are capable of doing, we remain unsure of our goals and we seem to be as discontented as ever… We are more powerful than ever, but have very little idea what to do with all that power. Worse still, humans seem to be more irresponsible than ever… Is there anything more dangerous than dissatisfied and irresponsible gods who don't know what they want?*" (Harari 2015).

Leo Tolstoy, the Russian author, wrote, as early as 1900: "*There can be only one permanent revolution—a moral one; the regeneration of the inner man. How is this revolution to take place? Nobody knows how it will take place in humanity, but every man feels it clearly in himself. And yet in our world everybody thinks of changing humanity, and nobody thinks of changing himself*" (Tolstoy, 1900). Fundamentally, a sustainable revolution begins at home, just like charity. We need to take responsibility for our actions.

Imagine the sign carried by the person following you around in a more improved, responsible consumer world. The sign now shows forests replanted by the company making the furniture, a new school for the children and an end of child labour. Factories with good pay and working fire escapes. Fresh water and unpolluted food around the mines. Clean rivers where fish can be caught and eaten. That's a sign you won't mind following you around. And that's what responsible consumerism is all about.

What we see here is the importance of a functioning vital society. And this is where we return to the work of Adam Smith. His concept of a *laissez-faire* economic model, where interference is reduced and the economy is free to develop and flourish was founded on the idea that each individual found their moral compass from a healthy society. Individuals then resonate with this moral society and their sympathies or judgements are tempered so that they operate under an invisible hand, making responsible decisions that guide the economy in the right direction.

However, I wish to set out a more rounded concept of the invisible hand. Here, society is in resonance with its landscape (the environment), and the individual is in resonance with both. From this resonance emerges the invisible hand that acts across all of our actions as individuals for the good of society and the environment. This is a radical approach but is found in indigenous thinking around the world, from the Andean thinking of *Buen Vivir*, through to the Ogiek people of Kenya. And it is the latter group that I wish to turn our attention to now.

VIII.5. The Ogiek people and the new, improved invisible hand

The Great Rift Valley is one of the most powerful demonstrations of active geology on the planet. Six thousand kilometres in length, it stretches from Lebanon in the north all the way to Mozambique in the south. It's what is called a *divergent plate boundary*, where two huge sections of the Earth's crust are slowly moving apart. Eventually, it will carve off a massive part of East Africa, forming a new continent, as water will rush in to fill the valley. This is how the Red Sea and the Atlantic Ocean formed. However, it will be a while before this new continent appears.

For now, the Rift Valley is at its most impressive in Kenya. The Mau Escarpment, some one hundred miles (150 km) north west of Nairobi, is over ten thousand feet (3000 metres) deep. And growing on the edge of this escarpment is the largest remaining near-continuous block of montane indigenous forest in East Africa, The Mau Forest Complex. The Mau forest occupies some ninety thousand hectares (900 km^2), and supplies some forty percent of the water for Kenya. Within this forest live an ancient tribe, the Ogiek people.

The Ogiek people are hunter gatherers, and some of the last indigenous forest dwellers on the planet. They represent the remnants of the original aboriginal inhabitants of Eastern Africa. There are currently around eighty thousand Ogiek in Kenya (Kenya National Bureau of Statistics, 2010), though possibly as few as five hundred of them speak their traditional language, *Akiek*. The name *Ogiek* literally means caretakers of the plants and the animals. By working with the forest, this tribe contributes to the maintenance of the vital water supply, representing one of the rare examples of humans actively contributing to the provision of essential ecosystem services.

Perhaps the most interesting aspect of Ogiek lifestyle is the central role that honey plays. It is said that in order to understand the Ogiek, you must understand the bee. Not only is this a comment on the physical, but also the metaphysical. For an organism truly embedded within its ecology can only be understood within that context. Another telling point is that there is no word for beekeeper in the Ogiek language. This underlines the fact that the bees are not subjugated, but, rather, are partners in a collaboration.

A young working member of the tribe will eat up to four pounds (two kilograms) of honey per day. They make beehives from hollowed-out tree trunks wrapped in cedar bark, carefully removed from trees in order not to damage them, and the council of elders determines how many beehives should be built. The hives are about five feet long and two feet wide and are placed some forty feet off the ground in order to deter the honey badgers (not the Kansas City Chiefs' safety, Tyrann Mathieu). Sometimes the hives will be hung from the tree using vines, deterring even the honey badger, but at higher altitudes with greater rainfall, the vines quickly rot.

The Mau forest ascends to a height of ten thousand feet. This means that the tribesmen must follow the bees across the seasons, as the bees shift location depending on seasonal patterns of flowering and rainfall. There is always something in flower across this varied landscape. Rather than migration, it is more like seasonal work where they can 'commute' to wherever the bees are. The tribe have an acute ecological knowledge, knowing which tree species flower at what time, and where. However, when the high forest is in flower, small groups of tribesmen may gather and hunt there for several days or even months.

Honey is collected and carried in honey barrels, constructed in the same way as the hives and carried on the back. Importantly, not all the combs are removed from a hive, ensuring that the bees continue honey production. As much as two hundred pounds (four hundred kilograms) of honey can be collected and brought to the family by a single tribesman each year, in addition to that consumed *en route*. If a family have suffered hardship, honey is shared within the community, as a form of social security. Honey is also used to preserve meat and in making an alcoholic beverage.

Honey not only provides food. It is central to the economy of these people. It is bartered for a wide range of things with local tribes such as the Maasai. Trade with these tribes also helps secure peace. Because honey gathering involves a very specific skill set and ecological intelligence, the other tribes rely on the Ogiek for a supply. It is also sold. Money can then be used to buy things not accessible through bartering.

At a societal level, honey plays a central role in the Ogiek culture. Honey water and honey wine are used in rituals of purification, marriages, new homes, births and peace offerings. Honey plays a role in ritual communication with the spirits of the dead, in order to resolve issues that were carried to the grave.

In addition to a wide range of herbal medicines, where medicinal herbs are protected by the council of elders, honey also plays a significant role in health treatments. Interestingly, herbal medicines are not taken only when someone is ill, but are added to almost every meal, representing a prevention strategy rather than merely a cure. All of these rituals involve the non-economic use of honey. This may seem like a bad business plan, but actually situates this golden nectar at the centre of society.

And so, we see the three arenas, economics, society and environment, as tied together with honey. It is honey that forms the bridges across which the invisible hand reaches out. Each arena informs and infuses the others, and pervading all is ecological intelligence and a sense of integrated decision-making.

By contributing to the bees' success, pollination is safeguarded, a crucial aspect of forest sustainability. All tree species are protected to ensure that the flowering cycle is intact, and the number of hives controlled. Again, this is where ecological intelligence comes into play.

In terms of governance, a council of elders makes decisions on all that happens within the community and the forest. Historically, there was no tribal chief. Rather it was a collective leadership of the most experienced members. This was the case for the majority of African tribes until colonization. The invading European nations insisted on speaking to a single representative during negotiations, and thus imprinted the Western concept of a leader onto the African tribes. However, this was an artefact of empire. It is interesting that during great times of crisis, for example the Second World War, Britain dissolved the usually partisan system of government and opposition, replacing it with a more collective approach, the Churchill war ministry.

The council of elders also direct the education of their people, focusing on sustainable practice and the governance of the landscape. For example, trees cannot be felled without the council being consulted and medicinal plants are similarly protected. Hunting and gathering also require approval. The forest is divided among family clans using natural features such as streams or hills. Each territory has sufficient trees of the right species, providing ample nectar for honey production.

Each clan has sole rights to place hives in their territories. The territorial approach to resource management delivers accountability, wherein the success of the clan depends on it caring for the natural resources within its territory. Rather than a free-for-all, exploitative approach, without consequences, here the adage *"you reap what you sow"* has real meaning. If you chop down the fruit tree to burn the wood, you won't have any fruit. Traditionally no cultivation is carried out either side of streams and rivers, protecting the water from soil erosion and pollution from fertilizers (Ottenberg, 1960).

The integrity of the forest is central to the success of the bees, and by caring for the forest, the community ensure that ecosystem functions such as clean water, stable soils and diversity are safeguarded, while avoiding soil erosion, eutrophication, flooding and ecosystem collapse. Yet the Ogiek are treated with disdain and disrespect by many of the neighbouring tribes who have embraced a more Western lifestyle. In the *Maa* language, spoken by the dominant pastoralist Maasai tribe of the area, the Ogiek are referred to as *Iltorobo* or *Dorobos*, meaning *'those without cattle'*. This is an insulting term, implicating that they are poor and incapable and must eat wild animals as they don't have the ability to tend herds. It also carries with it the idea that these people are in poverty. This is a typical attitude now seen in much of the Western, Northern development programmes, where the rich successful and enlightened generously stoop to give aid to the poor, failed, primitive peoples of the South.

Yet the role of the Ogiek and many other indigenous people in securing reliable water supply underpins the very survival of the pastoralists themselves during the dry season. This fact was highlighted in the recent IPBES (2019) report, where lands occupied by indigenous peoples were declining less rapidly than the rest of the planet. However, issues such as climate destabilization, air pollution and droughts and floods, although not caused by indigenous people, still impact on them. But these are not failed, primitive people. Rather, they are integrated within the Earth system.

While the Maasai refer, abusively, to the Ogiek as *"mere animals"* (Blackburn, 1970), in some respects this reflects on a different truth, that they are integrated within their ecosystems like all of the other life forms, living with ecological intelligence and tuned into their landscapes and communities. These 'animals' are true guardians of the forests they inhabit because they recognize themselves as one with the forest, and dependent upon it.

There is no place for Garrett Hardin's tragedy of the commons here. This is because there is a much greater invisible hand at work, far beyond what Smith envisioned, that recognizes the foundation of the environment first, society second, and economics third. You don't dig up your foundations to sell them in the open market. Rather, the economy is shaped by the landscape, and the landscape is seen as sovereign.

Hardin failed to grasp the importance of the invisible hand between landscape and society, and how it tempers our thinking. For in the real, integrated world of these people, the invisible hand is not invisible at all, if you are looking in the right places. There is no tragedy here, because a properly functioning society, embedded in a landscape, is empowered to make the right decisions. This is ecological empowerment, far more powerful than individual empowerment.

Empowering the individual is a common theme that often embraces individualism, independence and personalization. Berger and Neuhaus (1977) situated individual empowerment within the context of the community, where mediating structures, such as family and neighbourhood, could empower individuals. We suggest that the individual cannot be the unit of empowerment. From a post-development position, interdependence runs deep within indigenous philosophy. Husband (1995) writes *"In non-European cultures, the self-evident primacy of the individual in relation to the collective cannot be assumed."* Mbiti (1969) states *"I am what we are."* Ubuntu, a sub-Saharan African philosophy, can be summarized as the concept that no one can be self-sufficient and that interdependence is a reality for all (Nussbaum, 2003).

Other philosophical positions broaden the interactive net still further. *BuenVivir*, the Andean philosophy, stresses that well-being can only exist within a community, and the concept of community is expanded to include Nature (Gudynas, 2011). Wilks (2005) suggests that *"feminist ethicists have argued that our moral identities are located in and constructed through our caring relations with others."* Tronto (1993) argues that humans are not fully autonomous but interdependent and that relational autonomy involves mutual dependencies.

Post-modern thinking rejects an empirical, reductionist approach wherein the empowered community is built from empowered individuals (Meekosha and Shuttleworth, 2009). MacIntyre (1999) concluded that we do not have individual rights at our foundation, but that we are irreducibly social animals. He discussed the virtues of *acknowledged dependence,* contrasting this with the virtues of independence of Friedrich Nietzsche, who sought freedom from the imprisoning power of relationships in his novel, *Thus Spoke Zarathusthra* (MacIntyre, 1999).

Questions have been asked relating to community empowerment as an emergent entity. How can a modern community be empowered if they lack the fundamental connections of kinship and economic and emotional connectivity that existed in past communities? (Hennick et al., 2012).

Carl Rogers, the American psychologist, set out his theory of *actualization,* wherein he claimed that humans had one fundamental drive: to fulfil one's potential by becoming a fully functional person (Rogers, 1959). This is a very Western approach, focusing on the individual as the unit of organization, whereas in other cultures, the achievement of the group is valued far more highly than the achievement of any single person. Rogers also failed to incorporate the natural environment into his thinking. Rather than self-actualization, a truly integrated human would, I suggest, display *eco-actualization,* not optimizing for the individual but optimizing for the Earth system. Ecological empowerment is then defined as the liberation of this eco-actualizing tendency across all levels of societal and ecological organization. We see this emphasis in the Ogiek people, and in much of the post-development literature.

Table I shows how the honey economy actually covers a large number of the sustainable development goals (SDGs) recently developed by the UN. The Ogiek have been doing this stuff for at least a thousand years, and doubtless, much longer. When we examine the seventeen SDGs, we find that, with the possible exception of goal 7 (Clean energy), all of the goals are covered. Indeed Goal 7 is very much a recent challenge, and is only really an issue in industrialized societies.

It isn't that the Ogiek have sat down one day, scooping handfuls of honey out of a clay jar (which, by the way, they make themselves from clay within the forest), and decided to design a society that matched the SDGs. Rather, their way of life emerged as part of a greater whole, where they found out how to live as part of the Earth system. It isn't about listing one hundred and sixty-nine targets and two hundred and thirty-two indicators, as do the SDGs. It is about learning how to live as part of a team, where the environment is respected, society is accountable and aware, and the economy is in resonance with society and environment.

What results is an indigenous community that nurtures its landscape, and naturally does the right thing. All over the world, indigenous communities are integrated within their broader ecology, be it the Inuit and Sami of the Arctic, the Pila Nguru of the deserts of Western Australia or the Pumé people of the Venezuelan savannah. These are equitable societies, where resources are shared. But most fundamentally, activity (economics) is centred around a societal context, and society is rooted in the environment.

From each landscape emerges the cultural identity of the people. It must be thus, as each landscape has its own set of opportunities and challenges. Ecological intelligence is important to these people, and life is not as tough as you might think. Research has shown that the average hunter gatherer works for under six hours each day, compared to agricultural and industrial societies where the figure is almost nine hours each day (Sackett, 1996).

Table I: How the honey economy of the Ogiek contributes to the sustainable development goals

Goal	Impact of honey economy
1. End poverty	Honey used in bartering and sales
2. End hunger	Honey as a major food, easily stored and used in meat preservation
3. Ensure health and well-being	Important medicinal role. Honey also central to cultural and spiritual rituals
4. Quality education	Ecological intelligence lies at the heart of the tribe
5. Gender equality	Council of elders of mixed gender, giving equal representation
6. Water and sanitation for all	Clean rivers as outcomes of good forest management
7. Decent work for all	Every member of the tribe has an active role and continuously learn new skills
8. Infrastructure and innovation	The forest is home and ecological intelligence generates innovation
9. Reduce inequality	Equitable division of forest between families and sharing of honey if needed
10. Resilient and sustainable communities	Cohesive focus on environmental and social integrity
11. Sustainable consumption	Central to all of the tribe's activities
12. Stop climate change	Forest acts as carbon sink
13. Life in the water	Intact forest delivers clean water
14. Life on land	Sustainable conservation of habitats
15. Peace, justice and strong institutions	Council of elders
16. Partnerships for goals	Humans and environment work together as one system.

But what of us today? We live in multicultural societies, and often, the traditional culture of our landscape is lost. However, we can re-connect to these landscapes and allow them to shape our otherwise different cultural backgrounds. Indeed, the landscape may well be the most powerful common ground for people from very different cultural backgrounds to interact, because it reaches through to our fundamental sense of being.

Localism, in terms of eating the food that grows in our areas without intensive industrialized interference nor polytunnels covering our fields, is a good starting point. Eating seasonal foods is also helpful, as it ties us into the natural cycles.

Also, education that has a much larger proportion of time spent outdoors, involving landscape as a central aspect, would be beneficial. Building up our ecological intelligence, with wild camping and wilderness experiences, again would help to re-integrate us into our natural context. The Scandinavian nations refer to this as *Friluftsliv*, literally meaning *'free air life'*. Gelter (2000) defines it as a "*philosophical lifestyle based on experiences of the freedom in nature and the spiritual connectedness with the landscape*". Friluftsliv is not so much about outdoor pursuits, as about outdoor reflection.

What then is this new way of thinking? It lies at the heart of a transformation to re-integration. It involves recognizing that the Earth system is not only a superior intelligence, but that it is the only game in town. We have fought and conquered Nature, but in doing so we have humiliated our very saviour. We must stop and recognize this, finding our solution space within Nature rather than apart from it.

We have explored the characteristics of Nature and how it operates not at the individual level, nor at the level of one single species, but rather as a whole, where each participant must embrace sub-optimality for the greater good. This takes humility and teamwork. But it also involves emergence and the willingness to listen very carefully.

VIII.6. Lessons from the edge of the world: The St Kildan legacy

For a final case study, let us return to the tiny island archipelago that is St Kilda, home of the St Kilda wren. Surprisingly, the St Kilda archipelago shares much in common with the Ogiek landscape of Kenya. The islands that make up St Kilda are the remnants of a large volcano, part of a chain of volcanoes, located in an ancient rift valley that produced the Atlantic Ocean, very similar to the Great Rift Valley of East Africa.

It is thought that the original volcano formed some sixty million years ago, soon after the K/T mass extinction and around the time that primates first appear in the fossil record. As the continental plates separated, the crust was stretched and became very thin. Magma poured to the surface and, along the active rift, a chain of volcanoes was created. This is geology on a massive scale. Sixty million years later, the same basic process is occurring in East Africa.

Perched on the edge of the remnants of this ancient volcano, now eroded and just protruding above the ocean's surface as a series of small islands, lived a population of people who were hunter gatherers for most of their existence, much like the Ogiek. The St Kildans were not nomadic, not because they lived in a forest full of all that they needed, but because they were surrounded by an oft times hostile ocean, and had to make do with what they had. Indeed, there are no trees on the St Kilda archipelago. Trade was in the form of feathers, oil and sheep wool rather than honey. Their economy revolved around seabirds, not bees.

The people of St Kilda were self-sufficient, utilizing grasses, peat, sheep, fish and birds. Their houses were built of stone walls set three feet deep into the ground for insulation, and roofed with turf and thatch painted with pitch. These traditional structures are known as *blackhouses* and are found throughout the Scottish Highlands and Islands. The diet consisted of the many seabirds that migrated to the island each year, including gannet, puffin and fulmar, whereas seafood was relatively unimportant. Barley, potatoes and oats helped boost winter supplies, and bird eggs were also eaten. In 1764, it was recorded that each islander ate a staggering thirty-six eggs and eighteen birds each day.

In 1876 alone, the islanders caught ninety thousand puffins for food. The St Kildans also ate one hundred fulmars each per year (Fisher, 1952). But there were plenty of seabirds. Evidence of the sustainable nature of the bird harvest is clear: after 1930 when all humans left St Kilda, the bird populations did not increase. Oil from fulmars fuelled lamps. Seaweed and peat provided fuel, as there was no wood. Meat was air-dried or salted in small stone huts called *cleits*.

Even today, St Kilda has the world's largest colony of gannets and Europe's largest colonies of both puffins and fulmars. Indeed, St Kilda is Europe's foremost seabird colony, and a key seabird breeding station in the North Atlantic. The archipelago is one of the few locations in the world that has been awarded 'mixed' World Heritage Status, recognizing both its natural and cultural significance.

The islanders had no use for money until the nineteenth century. They lived in a feudal communalist system, where they paid the landlords, the Macleods from Skye, in feathers, oil and sheep wool. This was collected once a year, and islanders gradually filled a building (*the feather house*) with plumage over the previous twelve months.

No-one on the island owned property (hence the communalist tag) and decisions were made by the collective St Kilda parliament, consisting of all the adult males. This parliament met on the main street every weekday morning. There was no chairperson and no rules. The work plan for the day was decided at this meeting. Food was divided between families depending on the size of the family, a completely egalitarian system.

You already know that the humans were removed from this harsh environment in the Atlantic Ocean in 1930, bringing to an end perhaps ten thousand years of habitation. But the story that led to the collapse of this quite remarkable population is worth examining in more detail.

To understand the unique and ancient lineage of these islands, we need look no further than the sheep. Soay sheep (*Ovisaries*) are believed to be the only living relic of the earliest semi-domesticated sheep stretching back to the Neolithic times. The sheep lived on the uninhabited island of Soay, part of the archipelago, and wool gathering forays would occur several times a year. These primitive sheep do not form fleeces, but rather shed their hair, and so it must be plucked from the sheep by hand.

Martin Martin, the Scottish writer and traveller, visiting the island in 1697, noted that: "*The inhabitants of St Kilda are much happier than the generality of mankind, being almost the only people in the world who feel the sweetness of true liberty*" (Martin, 1697). He further noted that "*the voice of one is the voice of all the rest, they being all of a piece, their common interest uniting them firmly together*".

Macaulay (1764) refers to a state of "*happy ignorance, which renders them absolute strangers to those extravagant desires and endless pursuits, which keep the great and active world in a constant agitation*". In 1838, Lachlan MacLean wrote "*If St Kilda is not the Eutopia [sic] so long sought, where will it be found? Where is the land which has neither arms, money, care, physic, politics, nor taxes? That land is St Kilda*" (MacLean, 1838).

However, life on the archipelago was not always a utopia. The islanders were extremely susceptible to diseases brought from the mainland. In one terrible event in 1727, the population of one hundred and twenty was reduced to only thirty (four adults and twenty-six children according to Macaulay, 1764) by an outbreak of smallpox. The disease had arrived on the island on clothes returned from the Island of Harris after a St Kildan had died there.

Other diseases such as leprosy had similar devastating impacts. Infant mortality was extremely high, with many of the new-borns dying from tetanus. In 1852, thirty-six islanders migrated to Australia, but many of them succumbed to disease on the journey, far more than normal. This again pointed to their immune systems being unchallenged and unable to respond to even fairly innocuous infections. An aside here is that the Melbourne suburb of St Kilda, now home to a famous Australian Rules Football club who won the 1966 Grand Final, was named by the survivors.

In the 1860s, a new building material was introduced by the laird, tin roofing sheets. It seems hard to grasp even now, but this would contribute to the decline of an ancient civilization. The peat roofs took a huge amount of maintenance, but were very well insulated. The tin roofs were so much easier to construct, but were not well insulated. They also leaked in the driving rain and winter storms where winds could reach one hundred and sixty miles per hour.

The peat fires that were used to cook and to heat the stone cottages didn't release enough heat to survive the harsh winter temperatures under the new tin roofs, and so the islanders had to start importing coal. They didn't have any money, but the coal companies would not accept wool, feathers or oil as payment. As a result, the male islanders were forced into seeking paid work on the mainland. The population began to collapse.

Alongside these events, Christianity came to the island in the eighteenth century. Previously the islands had been the last stronghold of druidic belief. A puritanical minister, the Reverend John MacKay, arrived and instituted so many religious practices that there was insufficient time to catch enough birds for food. MacKay banned music, which the islanders had loved, and the weekly fishing trip on a Wednesday was replaced by a prayer meeting. He also introduced a hierarchy for the first time, appointing elders to report back to him about any sins they found out about (Steel, 1975).

This one man did terrible damage to these people and their ancient culture, replacing a joy in their natural world with a guilt of sins unrepented. At the same time, the soils were becoming less fertile, possibly because of the build-up of lead and zinc to toxic levels. The lead came from peat ashes and the zinc from both ash and bird carcasses, both of which were used as fertilizers. But over time, this began to inhibit crop growth.

Tourists began arriving in significant numbers in the nineteenth century and soon the islanders were making and selling things to them for money. Money brought inequality, as some families earned more than others. Slowly but surely, over a fifty-year period, the social structure that underpinned the community was eroded, as was the close relationship with Nature. The lure of the mainland, with all of its sophistication and civilization, became too strong a force to resist.

The first woman to leave the island was Christina MacQueen, in 1905. But it was her sister, Mary Gillies, whose death would hasten the end of the St Kildan population. Mary was pregnant, but appendicitis meant that she needed to be rushed to the mainland in February, 1930. The weather was terrible, and it took two weeks for her to be rescued. She died in Gartnavel Hospital in Glasgow along with her new-born baby, Annie.

The remaining islanders couldn't cope with this final blow, and petitioned to leave the island where they and their ancestors had lived for thousands of years. On 10th May, the remaining thirty-six islanders wrote a joint letter to William Adamson, the Secretary of State for Scotland, requesting evacuation. The original letter, an incredible piece of history, can be accessed at the National Records of Scotland (1930).

On the 29th August, 1930, all of the remaining islanders boarded a boat, the *Harebell*, and, along with their belongings, were relocated to the mainland. The ships log records the event in stark terms: "*Ship's log, 29/08/1930: At 5am this morning, HMS Harebell lowered her whaler. Embarked the inhabitants of St Kilda, 14 women, 13 men, 15 children…everyone on board. Hoisted boat. 08.02 hrs Weighed. Proceeded 11.5 knots*" (in Hunt, 2015). The previous day, the men had tied stones to the collars of their dogs and had thrown them into the sea, as they were not permitted to bring them with them, undoubtedly a traumatic moment for all who witnessed it.

Writing of this momentous event, Lieutenant Commander Pomfret, the surgeon on the *Harebell*, recorded that *"all the houses were locked and the people taken on board. Shortly afterward they were looking their last at St Kilda as the Harebell, quickly increasing speed, left the island a blur on the horizon. Contrary to expectations they had been very cheerful throughout, though obviously very tired, but with the first actual separation came the first signs of emotion, and men, women and children wept unrestrainedly as the last farewells were said"* (in Hunt, 2015).

Perhaps most difficult for the islanders was the fact that they were leaving their dead, buried relatives on the island. Christina McQueen, reflecting on this final exodus, poignantly wrote: *"May God forever guide the steps of those who, in waking or in sleeping, will ever hear the spirit voices of those they have left behind in* eileanmoghraigh! *(my lovely island)"* (Brown, 2010).

The last of this ancient lineage, Rachel Johnston, died in 2016 at the age of ninety-three. She had been eight years of age at the time of the great evacuation.

St Kilda has attracted huge attention, particularly in the eighteenth and nineteenth centuries, in terms of its idyllic way of life, its tight community and the fact that it was somehow apart from the restless, hurried and selfish existence on the mainland. Many books were written at this time, examining this remarkable place and people, including Martin Martin's *A Late Voyage to St Kilda, the Remotest of all the Hebrides* (1698), J. Sands' *Out of the world, or, life on St Kilda* (1876), Macaulauy's (1764) *The History of St Kilda*, and Seton's (1877) *St Kilda*.

A recurrent theme was the mix of the contentedness and social integrity of the St Kildans, juxta-positioned with the harshness and cruelty of the landscape. Edmond Burke wrote that *"Whatever is fitted in any sort to excite the ideas of pain, and danger, that is to say, whatever is in any sort terrible, or is conversant about terrible objects, or operates in a manner analogous to terror is a source of the sublime; that is, it is productive of the strongest emotion which the mind is capable of feeling"* (Burke, 1759).

Here he equated the terror of remote, rugged landscapes with a stimulation of the mind. The point being made had a more profound meaning than this, and finds its roots in the Enlightenment writings of Adam Smith, David Hume and Hugh Blair, who pointed towards social order emerging from individual sympathies or morality, that, in turn, were in resonance with the society within which they developed.

Thus, great emphasis was placed on education and social improvement, in order to promote the moral health of society. Philosophers then became concerned that such improvement could stagnate and level off, unless there was further stimulus. What Burke wrote about, and what became associated with St Kilda, was that the fear that some of the great landscapes such as St Kilda evoked within us would awaken us to a more profound understanding of ourselves, building social cohesion. This theme became central to the early literature on St Kilda (see Macdonald (2001) for an excellent exposition on this).

And so, we find St Kilda being held up as an example of how things could be, a celebration of humanity in a sublime setting, where happiness was not reliant upon earthly pleasures, and the society functioned in an ideal way, far from the sordid mainland. St Kilda came to represent the utopia sought after by the Enlightenment philosophers. No war, no weapons, no politics, just a people working at one with each other and their environment, as imposing and menacing as this environment was. Such an approach tended to exaggerate the perfection of the St Kildans, and later writers such as Connell (1887) appear to question the reality of this idealistic vision.

Yet time and again, the romantic symbolism of St Kilda has been returned to. On 4th August, 1999, member of the environmental protection group, Greenpeace, met in the main street on Hirta, the largest island of the St Kilda archipelago, and reconvened the Parliament for the first time since 1930, poignantly referencing the community meetings so vaunted in the old writings. The new Scottish parliament paid homage to St Kilda in its architectural design (Lorimer, 2002).

In whatever way you look at it, Burke was exploring a tripartite relationship, between environment, society and economics, attempting to extend the invisible hand to reach across all three arenas. Yet his work was completely ignored in terms of economics and the ultimate politico-economic expression of the Enlightenment, capitalism.

Unquestionably, the St Kildans had survived thousands of years on a tiny archipelago in extremely harsh conditions, and this could only be achieved by living sustainably and in social cohesion. However, the entire system collapsed when it became corrupted by the fruits of the mainland Enlightenment. From tin roofs to money and from tourism to tobacco, the blight of colonialism would ultimately destroy this ancient community.

Sadly, the Ogiek too are dwindling as their language becomes lost, pesticides used in surrounding farms kill their bees, and successive governments sell off the forest to timber companies.

So where does this leave us? Clearly, we cannot and should not pursue the utopian dream by taking to the local forest and building beehives, nor should we start eating gannets. There are seven billion of us. There are less than two million Northern gannets. That would be one gannet between three thousand five hundred of us. Sustainability requires a much greater transformation. However, the choices are both straightforward and easy qualitatively. They are straightforward because we have no other option, unless a crash landing attracts you. They are easy because the whole solution is already there, all around us, still working beautifully in spite of our assault on it. The biggest challenges are to recognize these points (our attitude) and to act upon then (our application). But to help us on our way, we have AI, that, with its worthy sidekick, the internet of things, can step forward as the means to these ends.

VIII.7. The Garden of Eden complex: How not to fix the world

Our mentors should not be the Ogiek or the St Kildans *per se*. We can learn from them of course, but the point is that their cultures emerged from their landscape. It is our partnership with the Earth system that paves the road to a sustainable future. We must ask only two questions: do our actions strengthen and resonate with the Earth system and do our actions strengthen our communities?

Central to everything is feedback. When you are dealing with a hugely complex entity such as the Earth system, it is akin to a black box, where you have inputs (human and natural) and outputs (natural). What happens within the black box is complex and mostly hidden amongst a myriad of connections, but what you can measure are the inputs and outputs. This is an emergent system, but there are some simple realities. What goes in impacts on what comes out the other end. Thus, in order to know what to put into the box, we need to listen very closely to what comes out. Then, we can adjust our input accordingly.

It is important to remember that basically every part of the Earth system has been impacted by us. Recently, explorers undertook the deepest-ever dive to the bottom of the Mariana Trench near the Philippines, the deepest point in the ocean. It was an extremely challenging dive, some seven miles (eleven kilometres) beneath the surface, deeper than Mount Everest is high. Only four people have ever been there. Fewer people have been there than have walked on the moon. They were excited about what they would find, and the trip didn't disappoint. Four new species of amphipods were discovered. Sadly, a plastic bag and some sweet wrappers were also discovered.

Given that atmospheric pollution, water pollution and land degradation impact globally, we do not have a perfect natural ecosystem to use as a reference point. However, we do have some fairly intact ecosystems to give us pointers.

It's not about us reverting to our *Homo habilis* traits. Rather, we need to stop doing what is clearly bad for the environment, and therefore bad for us, and then focus on helping Nature heal itself. This generally means reducing our interference rather than increasing it. As we have emphasized throughout this book, the Earth is a system, not a structure built of bricks. Reductionist, empiricist approaches don't work here, and that is possibly the most difficult thing for us to accept. We need to recognize what a system is, and focus on listening. Nature has the ability to recover, as it has shown through time immemorial, but we need to let it be.

Of course, we still need to live, and so does everything else. It is not about leaving no footprint in the sand, but it is about not racing across the beach in a Sherman tank (or, for that matter, a Soviet T-34. Other tanks are available, but equally not encouraged). It should not be a silent zone, in some cathedral or library. We are allowed to make some noise, just not too much. Indeed, noise is important. As predators we have an important role in controlling populations of our prey. As omnivores, we also help bring equilibrium to other layers in the food pyramid. How much noise and what type of noise? This comes back to the intermediate disturbance hypothesis. And our AI feedback will let us know if we are being appropriately noisy.

VIII.8. The three cornerstones: Diversity, resilience and integration

Provided that society and the environment are diverse, resilient, integrated and functional, then all is good. Diversity brings functional redundancy, meaning that if one species becomes extinct, the whole system can keep going. It also brings an array of response to challenges. If we were all the same, we would not have a range of ideas to tackle new issues.

Diversity operates at two levels: population diversity and species diversity. Population diversity is the diversity within one species in a particular location. It is a crucial characteristic of a healthy population. This is why inbreeding is such a problem. Inbreeding leads to harmful mutations in the DNA being concentrated in the offspring. Single organisms usually have some harmful errors in their DNA, but these are often masked by healthy DNA. However, closely related organisms share much of their DNA, particularly if they have the same parents. If two siblings mate, the dangerous DNA becomes concentrated in the offspring, leading to devastating developmental issues. A clear example of this arose in Nueva Germania in Paraguay.

Nueva Germania was an Aryan colony formed by Bernhard Förster and Elizabeth Förster-Nietzsche, the sister of Friedrich Nietzsche. Bernhard Förster was a war hero, awarded the Iron Cross, and was an agitator against Jewish power in Germany. It is important to note that this was before the birth of Adolf Hitler. Hitler did not invent anti-Semitism. It was alive and well in Germany and many other countries long before him. Friedrich Nietzsche was strongly opposed to anti-Semitism and wrote a book criticising Wagner, entitled *Human all Too Human* (Nietzsche, 1986). His later book, *Twilight of the Idols: The Antichrist* (Nietzsche, 1990) was a much more robust attack.

Both Elizabeth and Bernhard had been heavily influenced by Richard Wagner's book, *Religion and Art* (1880), that called for the colonization of South America by Germany. They took up the call and in 1886 founded their colony, based on German patriotism, Christianity and vegetarianism. Bernhard would ride around on a white horse, inspecting progress.

The colony quickly collapsed. Germans found the tropical heat unbearable, fell out with the locals and with Elizabeth the founder, who quit in 1893. Her husband had already poisoned himself in 1889. The colony only survived by growing large quantities of a caffeine-rich plant called *Yerba mate*, that was sold to make mate tea for significant profit. The Aryan dream drowned in a caffeine buzz. It had failed in its efforts of purification and rebirth of the human race through Aryan breeding.

Worse was to follow. The remnant of German families had a small amount of genetic diversity, that reduced further with inbreeding. The attempt to achieve a pure race led to anatomical and physiological mutations becoming expressed, and limb deformities, heart issues and other deformations.

This has been seen in many small communities and in royal lineages, both of which often have small gene pools. The most famous of these was the *Hapsburg lip*, named because of its increasing occurrence within the Royal House of Hapsburg. The Hapsburg monarchs ruled across Austria, Bohemia, Hungary, Croatia, Galicia, Portugal and Spain, and were highly inbred. The Hapsburg lip is a mandibular prognathism, where the lower jaw outgrows the upper jaw. This prognathism also occurred among the deformations within Nueva Germania. The point is that diversity in the gene pool of a population is essential for maintaining healthy functional communities.

The second important type of diversity is *species diversity*. The more species you have in an ecosystem, the better it is. It provides resilience, in that if some of the species are lost, others will be able to replace them and continue the functioning of the ecosystem. Having spare species is a great strength. Also, any two species do things slightly differently even if they both eat grass or chew nuts. These differences provide variety and resilience in times of stress.

As well as diversity and resilience, integration is an essential property of a functioning ecosystem. Integration circles around feedback. As we've already noted, feedback is one of the four key properties of any system, biological or physical. Integration means connectivity, both to your own species (through healthy societies), other species and the landscape. We are so unintegrated, so separate from the conversations that matter, and thus integration is one of the great challenges. Fortunately, integration is the business of AI and the internet of things, and the new technology can be an enabler, as we shall see shortly.

Finally, successful ecosystems are functional. They produce outputs that strengthen their own and the broader biosphere's existence. Clean water, clean air, a balanced chemistry, appropriate fluxes and flows all stand testimony to a functioning ecosystem. This functionality also includes the ability to recover from catastrophic disasters, to pick themselves up and recover diversity when it is lost.

Primary succession is where newly formed land, such as on a sand spit or a new volcanic island, becomes gradually inhabited by life, with communities transitioning until they form a complexed, stable ecosystem. In the case of sand spits, this is usually a forest. It can take one hundred and fifty years for this process to complete. However, often, a massive catastrophe, such as a fire or a volcanic eruption, can wipe out most of an ecosystem. The power to recover (an inherent and amazing property of ecosystems) is called *secondary succession*. This is self-healing and is one of the awesome properties of the Earth system.

I remember visiting the devastated region around Mount St Helens in Washington State in North West USA. We drove for miles through forest that had been flattened by superheated winds at speeds of two hundred and fifty miles per hour, released from the massive eruption at 8:32 am on 18th May, 1980. It was a scene of complete devastation, of a scale that shook you to the core. I've never seen anything like it. Two hundred and forty square miles of formerly pristine ecology had been turned into a haunting, grey apocalypse. The heat, power and voraciousness of this explosion, one thousand six hundred times the size of the nuclear bomb dropped on Hiroshima, had wiped out everything.

The blast had come from the northern side of the volcano, and all the tree stumps, their leaves turned to dust, lay, like a myriad of compass needles after an explosion at a compass factory, pointing North. Yet within a year, sprouting lupins could be seen growing through the ash, and gradually, Nature reclaimed its place. In time, a rich volcanic soil would form and become a utopian substrate in which to grow, packed with nutrients, and with good water-holding capacity.

The ability to recover and rebuild is the sign of a functional system. Such regeneration is one of Nature's great powers. It also gives hope. If we reduce the pressure on the natural system and, instead, switch sides to join forces with it, then we can also be regenerated as societies. This is because our species was once part of the Earth system. This is very much at the centre of post-development thinking, where humanity can be reborn through a re-connection with the Earth system. This book positions AI as the portal to this reconnection, and now it is time to look in detail at how this can happen.

VIII.9. The central role of AI in feedback: Shaping our new world

As we have stressed, real-time feedback is essential if we are to re-integrate into the Earth System. This feedback works on numerous levels. Firstly, it informs us of the outputs of the black box. Secondly it alerts us to the impacts of our inputs into this box. Thirdly it acts as an invisible hand between the environment and society, feeding into our ethical decision-making and making us aware of the natural reality.

This is important in terms of accountability and in re-enforcing the message that what we do has meaning, positively or negatively. It creates its own positive feedback loop, wherein once we see what difference we can make, it prompts us to make greater efforts. Also, it forms a reward system, if properly designed. Another aspect of this feedback is that it has the potential to strengthen society by building a team mentality. Let's consider how all this can work practically. We will examine this at three levels: *micro-feedback*, *meso-feedback* and *macro-feedback*.

Micro-feedback is a limited set of measurements, fed back to an individual, relating to our lifestyles and property, such as a house, a car, or what we buy. If we take a house as an example, this could include water use, electricity, heating, air conditioning or how eco-friendly our gardens are (in terms of the proportion of pollinator-friendly plants or indigenous species present, for example). Smart houses are already in existence, but it is the integration of this feedback that could be transformative. I envisage this as a panel somewhere in the house, that changes colour depending on how the ecological footprint in the house compares to the previous week. Perhaps a deep blue could represent improved environmentally friendly resource use, while red could represent an increase in environmentally damaging resource use. By displaying this summary of the house footprint as a colour scheme, all the members of the household can become accountable for their activities. A detailed breakdown could also be accessed if required, showing each component of the overall assessment, and providing an historical record of changes through time.

The advantage of a colour panel is that you don't need to plough through lots of data. Instead, you have a visual summary, easily understood by everyone, but that can be interrogated when and if required. For those who are visually impaired, an audio output can be added. The internet of things provides the data, and AI can analyze and display the outcomes. This data can also be combined with other data related to our other activities.

Our cars can have similar panels, reporting on the efficiency of the car (in terms of engine performance and tyre pressure etc.), our driving (in terms of fuel efficiency) and alternative approaches to our travel plans (including public transport). Again, further details can be gained, but the initial overall summary provides us with a nudge in terms of improving our travel footprints.

This analysis, anonymized, can be further examined in combination with other individuals in a community, county or country, providing local and federal government with an essential insight into population trends and what is needed. For example, car use data could provide important awareness in terms of where to place bus routes, park and ride schemes, timetables, cycle paths and so on. House data could highlight locations that may require government intervention in terms of improved insulation.

Lower rates and tax reductions could be used to incentivise individuals and families. One such approach is *personal carbon trading* (PCT). PCT is an umbrella terms covering a range of ideas, all related to cap-and-trade, where the individual takes responsibility for their management of domestic, vehicular and aviation energy management. In general terms, each individual is given an allocation of carbon (a cap). If they exceed this, they need to buy carbon units from someone else (a trade). Thus, someone using less than their allocated carbon each year can sell their unused credit for cash. Each year, the allowance for each person decreases, leading to an overall reduction in emissions year by year. Families who live low carbon lifestyles can profit from their actions, thus incentivizing greener living.

The idea has been around for fifteen years, but difficulties in implementing it have prevented it from being applied. Other issues remain in terms of equality. Many poor families live in energy-inefficient accommodation, often owned by a landlord, and have limited ability to improve the energy efficiency of their properties. They may have less efficient vehicles because they cannot afford newer, more environmentally friendly alternatives.

Giving out equal amounts of carbon credits to everyone may disadvantage those less well off. The rich can afford to buy credits off the poor, who may then sell their credits rather than heat their homes, leading to their dependent family members suffering, and increasing *fuel poverty* (where a family cannot afford to keep adequately warm). Also, in any country, different areas have different climates, and those with colder winter temperatures will use more domestic energy. Families in more rural areas could well need to drive longer distances, with less public transport options, than those in large cities. Lockwood (2010) has written a good overview of these and other issues relating to PCT.

But there is a deeper problem relating to PCTs, and this impacts on all similar schemes. It is not all about carbon. Carbon is an issue, undoubtedly, but the problems facing us are greater than just carbon. We've mentioned them throughout this book. Soil erosion, soil salinization, eutrophication, habitat destruction, water use and pollution from mining and manufacturing all need to be addressed. And there are environmental trade-offs within these. For example, if we look at energy production in terms of the water use per megawatt of energy produced (Table II), we can see that open-loop nuclear power plants use vast amounts more water than do oil refining or petroleum extraction. Corn ethanol and soybean biodiesel are another order of magnitude worse. Thus, the water footprint of low carbon emission energy is vast.

In terms of biofuels, most biofuel crops are extremely thirsty, requiring huge amounts of irrigation. In many countries, this creates a significant problem, as water is scarce. Also, land used for biofuels cannot be used for food production, meaning that food production on the remaining land must be intensified, further damaging the planet. Because the financial incentives for biofuel production can be high, there is greater pressure to convert forested areas into fields for growing biofuel crops.

Finally, much of the fertile land in many countries is in the form of small-holdings, where families practice subsistence farming. An example is the *shamba* system in Africa. For biofuel production to work, large tracts of land are needed, and this would mean the displacement of local people from their small-holdings. International companies have been shown to use coercion and undue influence in obtaining this land (Hikmany et al., 2015).

Overall, energy generation accounts for forty four percent of the water used in the EU, compared to twenty four percent for agriculture and twenty one percent for public consumption (Koulouri and Moccia, 2014). Thus, the water footprint for energy is huge.

Table II: Water footprints of different methods of power generation (Dominguez-Faus et al., 2009; Koulouri and Moccia, 2014)

Energy generation method	Litres of water per MWh produced
Soybean biodiesel	13900000-27900000
Corn ethanol	2270000-8670000
Nuclear (open-loop cooling)	94600-227100
Supercritical pulverized coal	1700
Nuclear (closed-loop cooling)	950
Gas power plant	700
Oil shale surface retort	680
Oil refining	80-150
Petroleum extraction	40
Geothermal	20
Wind	0

As we have seen, mining and extraction of rare earth metals, used in wind turbines and much modern technology, produce highly toxic waste and are toxic themselves (Zhang et al., 2000; Pagano et al., 2015). Lithium mining, essential for the batteries in electric vehicles, is disrupting indigenous land rights in many areas (Babidge, 2016). These indigenous people tend to suffer rather than gain from mining (Wanger, 2011; Revette, 2017; Humphreys Bebbington, 2013; Hancock et al., 2017). Solar panel production and maintenance produces highly toxic waste, including cadmium, sulphur hexafluoride, arsenic and iso-propanol (Aman et al., 2015).

Yet we are so consumed with carbon, that we fail to assess the complete ecological footprint, covering all of the environmental damage. Social footprints are also not assessed. This one-dimensional approach to the disastrous problems facing us will fail. It resembles the crew of a small boat, that has five holes in it, all focusing on fixing one of the holes. By the time that hole is fixed, the boat will have sunk. We can't afford this fixation with carbon. The jolly green giant may have increasingly smaller carbon feet, but the shadow of this giant, in terms of ecological damage done by this shift to 'green' energy, must be factored in.

More fundamentally, this means that energy production is not the only issue. While the ecological footprint of production is important, so too is the impact of increased energy flux throughout the biosphere. Broadening this concept further, it is the ecological footprint of our lifestyles that is important. This brings us to the society-economy nexus, because the goods that we purchase all have an ecological footprint, and we must be accountable for this, as must the businesses and designers who determine what these goods are made of. Here we return to the thorny issue of supply chains. And AI can provide the solution here.

Every item that we purchase should have an ecological footprint barcode on it. By swiping this with a hand-held device we could download the complete ecological impact of the product, from mining and fabrication of the raw materials, water used in the production of the product, land use, social issues, transport costs, and all of the other aspects of production involved.

This is particularly relevant for food items, but also for everything else. For example, did you know that a cotton t-shirt requires two thousand five hundred litres of water to make, whereas a litre of apple juice requires one thousand one hundred and forty litres of water to produce? A kilogram of beef requires fifteen thousand four hundred litres of water, and a cup of coffee requires one hundred and thirty litres of water. Interestingly, tea only requires thirty litres per cup. A litre of wine requires six hundred and fifty litres of water while a litre of beer requires two hundred and ninety-six litres of water (Mekonnen and Hoekstra, 2010; 2011; 2012).

By being able to download all of this data from our shopping items, an AI system could assimilate this with our domestic, vehicular and other resource consumption, feeding into the colour panel at home, giving us a combined colour summary of how we are doing, and, when needed, breaking this down into more detailed analysis. Shops, suppliers and designers would all use similar barcodes, and be informed of exactly what the options are.

If we are going to implement any form of cap-and-trade system, it should bring together a much more sophisticated personal ecological trading system (PETS!), wherein caps are placed on all ecological damage, rather than merely focusing on carbon. However, the same issues will still arise as identified for PCT, and a more improved system that addresses these issues will need to be developed.

AI can be central to this, allowing sophisticated metrics to be derived that bring together both our consumption patterns and our social situatedness, wherein everything is contextualized. Furthermore, this can operate at so many other levels. Each business would be forced to display its consumption panel in its front window, highlighting its progress in reducing its own ecological footprint. Designers would also have their records available, in terms of the footprints associated with their designs.

Government procurement can be analyzed and driven in a similar way. Indeed, AI can be put in charge of procurement, using the analysis of supply chains as a basis for decision-making. It is essential that governments do this, to prevent hypocrisy. We need to reach a situation where we are not discussing expense scandals of members of our parliaments, but ecological footprints.

Political parties should also be required to cost their policies in terms of ecological footprints during elections, so that we can vote on ecological consequences of political campaigns. What will a given political agenda deliver in terms of sustainability? This needs to be clearly stated in order to properly prioritize planetary health and safety. Again, this comes back to the issue of accountability.

Individual activities need to be situated within a social context, and social wellbeing and empowerment are rooted in cohesive and inclusive social policy. This is what we call *meso-feedback*. Functional societies are represented by functional communities at a local level. As populations have shifted from rural to urban contexts, the challenge of maintaining functional communities has increased.

Even in rural settings, villages have become dormitories, where people sleep, but commute to larger cities, spending little time in their local community. In more picturesque regions such as the Scottish Highlands, many of the properties in small rural communities are second homes, and these are only occupied for a few weeks a year, contributing little to the communities and pricing local people, who have an attachment to the landscape, out of the housing market.

Yet healthy, functional, located communities are essential if we are to realize our salvation from the ecological disaster that is emerging. And this is where meso-feedback, at the level of the community, comes to the fore, connecting people as a group with their landscapes, and creating an identity beyond the individual.

Adam Smith recognized the importance of our situatedness within a thriving society, and realized that this would provide the invisible hand that would temper capitalist excesses. Unfortunately, the free market economy ended up damaging society, with its neo-liberal, individualistic politics overwhelming the social collective that still persists in Andean, Asian and African societies. And the problem with this shift in thinking was that it also created a chasm between society and landscape.

AI can play a huge role in rebuilding society, particularly in terms of bringing communities together around a sustainable future. How could this happen? Once again it comes down to channels of communication. Having established a flow of environmental data for individuals, allowing accountability and responsibility to be taken, so, too, a collective responsibility can be promoted by pooling of data across a community. This can engender positive social action, in terms of working together to reduce footprints. This isn't a naming and shaming approach, highlighting the worst and best on the block. Data would be anonymised and pooled, providing an insight into the collective contribution to reducing our ecological frameworks.

Communities that make the most impressive progress could be awarded by local authorities. Communities could proactively work together in terms of decision-making, developing their own projects, all of which could feed back into their community scores. For example, a stream running through a village could be monitored. A meadow could be community-owned and monitored. The village school, nursery and playgroup could become sustainability hubs.

The local shop could supply local produce, with low ecological footprints. The vision here is not some kind of hippy commune, but rather a community taking responsibility for its actions and addressing the greatest challenge the human population has ever faced: our re-integration into the Earth system. This isn't a bohemian alternative, but rather an essential step towards restoring the diversity and functionality that our actions have damaged so drastically. The stakes are high, and society must function to deliver the transition required. Incentives can also operate at a community level, rather than at the individualistic cap-and-trade level of PCTs.

In order to empower communities, the benefits must be at community level, drawing people together in a team. Green bonuses, such as reducing the rates level for the community, may also help. Equality issues raised concerning PCTs can be averted by focusing on the community level, where success is defined as how well the group are doing, rather than how well an individual is doing. The road sign on the way in to the village could indicate a star rating, or a colour panel. Good practice could be identified and communicated through the internet. Different age groups could have different roles and children growing up in such a context would surely benefit hugely.

But what about cities? Today, fifty five percent of the human population live in urban environments, and this is expected to rise to sixty eight percent by 2050. Patrick Geddes, the town planner and social engineer, spent much of his life regenerating some of the most degenerate urban environments in the world. His principle was that culture and Nature were unified and interdependent, and thus he continuously designed green space into previously built environments, famously emphasising the need for engaging hands, hearts and minds. He envisaged people working in these plots of greenspace, and by so doing, reconnecting to Nature.

This in turn would transform society, by reconnecting humanity to the landscape. Geddes believed that culture was about place, and that humans needed to be rooted within their landscapes if they were to blossom. His work was inspired by the French sociologist Frederic Le Play's triad of *'Lieu, Travail, Famille'* (place, work and family), and would later be celebrated in the *Territorialist School* in Italy, led by Alberto Magnaghi. Magnaghi emphasised the importance of marrying three elements: enhanced environmental quality, self-reliance combined with local self-governance and provision of the fundamental human requirements (by no means only material).

Cities need redesign, in order to incorporate Nature. Sheffield in England represents a fascinating example. Here, they have connected up the parks so that they run continuously from the centre of Sheffield out into the Peak District National Park that surrounds the city, acting as natural corridors. The green spokes allow people in the city to feel connected, physically and metaphysically, to the great natural world beyond. Seventy two percent of Sheffield is green space. There are one hundred and fifty woodlands and fifty public parks within Sheffield. Over forty four percent of Sheffield residents live within a five-minute walk of a wood and half the city's population live within fifteen minutes of the open countryside. Such connectivity is very much what is needed in all urban environments.

Other possibilities include replacing non-fruiting plants with fruiting plants in towns, and replacing non-native plants with native species, thus encouraging native animal species. Roof gardens and balcony gardens should be encouraged. In order to increase Nature-based activities, community stores could be set up on each street, selling local produce that has been grown, made or crafted by residents. Schools can be heavily involved in this, as can residents who are retired or not working for some other reason.

The architectural historian, Volker Welter, discussing the importance of sociological co-operation, wrote "*For Geddes, conflicts arise not between classes but between occupational groups and the environment. As the aim is to adjust the whole city to the environment, cooperation among citizens becomes not only a viable option but a necessity*" (Welter, 2002). In other words, the meso-feedback level is extremely important in cities, and the challenge is greater here, as an urban environment can feel far removed from Nature. However, for a city to function as a society, the society must be able to access Nature in some form in order to maintain its link with the Earth system in a more than theoretical way.

John Law, Michel Callon and Bruno Latour stressed that humans, technology and Nature are one, and that humans cannot be seen in isolation from what makes them purposeful (Law, 1987; Callon, 1999; Latour, 2005). Humans and non-humans are intermeshed, according to their *actor-network theory*, and there is a constant shifting network of relationships. We may deny these linkages, but they are there. In many ways, the city has replaced the ecosystem for urban humans, yet it is clearly not a replicate.

Attempts to create an ecosystem surrogate in concrete and steel will only fail. It is essential that people interact with the real thing and in a Geddesian, meaningful way, emphasising a celebration of folk and place. Cities are isolated, non-integrated spaces, probably stemming from their design towards military protection, industrial efficiency and tight work/sleep cycles. Instead, we need to re-integrate the city with the ecosystem, thus, reconnecting the urban population with its ecosystem identity, so that the ecological context is an essential part of urban living.

The meso-level is an essential element of any transformation. Society ultimately has the key role in all of this, both in terms of transforming our individual attitudes, improving mental and physical health, empowering and enabling change and shaping economics to deliver healing rather than abuse to the environment. And AI has a central role in promoting cohesion through shared accountability and encouragement.

The bigger picture is so important. This is called *macro-feedback*. We can become so immersed in our own challenges and difficulties, particularly in terms of sustainable living, and can feel discouraged. Yet we are entering a time like never before in terms of being able to visualize planetary functionality. Remote sensing from a myriad of satellites and weather stations, connected to the internet of things, along with apps on our mobile devices, are already gathering immense amounts of data on planetary health, as we saw in Section VII.3. By bringing this data together and analysing it, AI can provide us with this big picture.

Trends, challenges, successes, good practice, all at a global level, can inform us locally. It was Patrick Geddes, who developed the concept of *think global, act local*. He explained it thus: "*'Local character' is thus no mere accidental old-world quaintness, as its mimics think and say. It is attained only in the course of adequate grasp and treatment of the whole environment, and in active sympathy with the essential and characteristic life of the place concerned*" (Geddes, 1915).

The power of AI lies in being able to assimilate data from sources never before available. This can allow us to grasp global issues in a completely new light. AI can deliver insights into planetary function thanks to the incredible array of technology that allows us to listen and hear the very breath of our planet. The internet of things infuses the Earth system, and by channelling this information into the unprecedented analytical power of AI, we can have our finger firmly on the pulse of our planet. The cicada hunt app is a good example, where all you have to do is turn on your mobile phone and cycle around, and the bioacoustic smartphone application listens for an endangered species of cricket, locating it within a forest (Moran et al., 2014; Jepson and Ladle, 2015). We need lots more of these in order to build further communication between ourselves and our environments globally.

By developing such technology, we can fill in gaps in our feedback, and all of this data can be channelled into the analytical powerhouse that is AI, with the resulting data-rich analysis available for all of us. National and regional levels can also be examined. This will allow us to challenge our governments and our international organizations, and reward us with tangible indicators of improvement in Earth system health, while reminding us of the challenges remaining. This macro-feedback will further bind society with environment, forming an important tool in the re-integration of the human race with the Earth system.

In addition to feedback on supply chains, it is important to know where the products we buy end up after use. Disturbing evidence has recently emerged that single use plastics in the UK, sent for recycling, have ended up being dumped or burned in Malaysia, causing significant public health issues. This is a disgusting example of externalizing pollution. Local councils have to be held accountable for this, or faith in such recycling will be lost among the population. Designers have responsibility to use materials that can be recycled, or to design products that can be re-conditioned. The best way to reduce the risks of huge amounts of waste returning to the environment is to reduce the production of the waste in the first place.

Furthermore, just because waste is recyclable, such as vegetable scraps or wood, care must be taken not to overload natural systems with excessive amounts. The Earth system works on a meagre diet of energy, water and nutrients. Pouring huge amounts of any of these back into the system is likely to have significantly destabilizing consequences. The pollution released from waste is a significant threat to the planet, needing urgent and honest efforts to address.

The recyclability of products should be included in the supply chain barcodes, so that consumers can know what the potential issues at the end of the product's life could be. Retailers, manufacturers and designers should be answerable for decisions relating to the stocking, creation and design of their products, and take responsibility for greening their products, not only in terms of their creation, but in their ultimate demise. The pharaohs spent their lives designing their journey to the afterlife, building their pyramid graves, and so designers should be primarily focused on reducing the damage of their products at the end of their useful existence.

What damage will this design cause to the environment and to society? What good will this design deliver to the environment and to society? These are the only questions that should inform the design brief. This demands a complete inversion in priorities. Instead of economic and commercial interests leading the way, where society is merely a market to be manipulated, and the environment just a source of raw materials and a dumping ground for waste, the sovereignty of a functional environment and the empowerment of a cohesive society must dominate. And the algorithms that lie at the heart of AI must reflect this. That being the case, many of the fears of AI becoming an horrific extrapolation of our own greed and destructiveness will remain in the realms of science fiction, while our actual technology will lead us on the path to our own redemption as we re-integrate into the Earth system.

Microsoft (2018) has stated that AI should be human-centric, serving the values and rights of the humans, developed over hundreds of years. This book strongly disagrees with the position taken by Microsoft here. By being human-centric alone, AI will fail entirely to sustain our environment, society and economics. It is this human-centric approach that has placed us in the position we find ourselves in today.

Horace (2002, Book III, 6, l.1) the Roman poet and son of a freed slave, wrote: "*For the sins of the fathers you, though guiltless, must suffer*". These sins, or errors in human thinking have led to the critical situation we now find ourselves in, with environmental damage reaching levels where, experts agree, the very future of humanity is threatened. If AI were human-centric, it would only hasten our demise, accelerating the destruction of the life support systems that we, unconsciously, are dependent upon.

Fields such as biomimicry focus on transferring biological structures and functions into the human domain for our benefit, and are, ultimately, human-centric. This book argues that this only reinforces the fundamental mistakes that have led us to where we are, stripping away context and blindly continuing down the broad road that will lead to destruction.

In order to stop future generations, though guiltless, from suffering, we must take a different path, wherein AI re-connects us with our ecology, and where the invisible hand not only operates for and within society, but, rather that it stretches across the three arenas, being eco-centric, not human-centric. Only by doing this can we contextualize our existence within that of the biosphere, and thus find the sustainable path forward. This path can be constructed, and AI can be the arbitrator and catalyst for the changes that are necessary, but not if it is human-centric and directed towards anthropogenic goals.

So that is the vision. AI can help save the planet by mentoring us and nudging us towards the better life, *buenvivir*. We don't need to wait until AGI is developed. The internet of things and current AI technology already offer all that we need. Next, we examine the barriers that stand in the way of this path, and what we need to do to overcome these.

SECTION IX Barriers to Change

"As long as the people of your culture are convinced that the world belongs to them and that their divinely-appointed destiny is to conquer and rule it, then they are of course going to go on acting the way they've been acting for the past ten thousand years… You can't change these things with laws. You must change people's minds. And you can't just root out a harmful complex of ideas and leave a void behind; you have to give people something that is as meaningful as what they've lost – something that makes better sense than the old horror of Man Supreme, wiping out everything on the planet that doesn't serve his needs directly or indirectly."

–Daniel Quinn, *Ismael: An Adventure of Mind and Spirit* (Quinn 1992)

Welcome to the penultimate section. We need to change. There is a path to a better way, and the technology already exists to facilitate our journey, but we must turn off the highway to hell and this has proved difficult. So, what is stopping us? The evidence of ecological decline is all around. Even if you don't 'believe' in anthropologically caused climate destabilization, and there are still people who don't, you can't deny the huge increase in air pollution leading to soaring rates of asthma, the spiralling rates of skin cancer, the plummeting number of species, habitat destruction, the disappearance of our wetlands, the rising ocean acidification, the dwindling resource levels, the exponential increase in plastic in the oceans, the rapid expansion of soil salinization and soil erosion, the expanding dead zones in the oceans, the collapse of our coral reefs, the escalating gap between rich and poor, the rise in political extremism and increasing social unrest. The evidence is all around us. And these things are all getting worse. So, it's not as if we don't have the information. Let's explore what prevents us from changing direction. There are three sets of barriers: philosophical, structural and psychological. Barriers are not limits. Limits can't be overcome, barriers can.

IX.1. The five philosophical barriers

The first philosophical barrier is that we don't take Nature seriously. The birds, the bees, pets, weeds and worms – at worst it is a case of Nature raw in tooth and claw, and at best, they live in our houses, but at no time do we view the organisms with whom we share the world as part of something bigger, the Earth system, that, in real terms, runs the planet. We are the self-proclaimed wise ones, and there is little or no room for, or acknowledgement of, the intricate, complex and interactive world beyond.

And this is a significant problem. Without such acknowledgement, we are a great distance from recognizing that the only way to build a sustainable future is within the Earth system, not apart from it. This is why biomimicry cannot work, because it shouldn't be about mimicking Nature, nor should it be about constructing some human solution with bits of Nature transferred into it. You can't build a parallel technological Earth system, and this is why weak sustainability cannot work. Natural capital cannot be substituted with economic capital. It doesn't work like that. And this comes down to the second barrier.

The second philosophical problem is that we fail to embrace emergence. Emergence is a central property of any system, and, quite definitely, of the Earth system. Yet emergence is something that modern, Western thinking struggles with. We like to think of the world as built from small building blocks, and we can add or take away blocks in order to change it.

This is the reductionist philosophy that has dominated scientific thinking since the times of Francis Bacon in the sixteenth century. We see it in weak sustainability, in genetic engineering, in conservation biology and in biological pest control (such as the cane toad), where we think we can just substitute bits of the system to solve problems, what I call the pick-and-mix approach. Newtonian physics relies on similar foundations, as, more generally, does the scientific method.

This reductionist thinking paves the way for intervention by us. It resonates with our *Homo habilis* compulsion. If we can understand what something is built of, then we can use this knowledge to manipulate things. Take genetic engineering. Once we accept that the gene is the unit of selection and the building block, then it is a simple, logical step to add or subtract genes in order to build a new organism.

Yet complex systems do not work like this. There are so many interactions that we cannot predict what will happen. Take the introduction of herbicide resistance to plants. Around twenty years ago, crop plants were genetically engineered to resist glyphosate, a weed killer. This meant that glyphosate could be sprayed at higher concentrations, killing weeds more effectively but not the crops. By 2014, ninety percent of cotton, corn and soybean grown in the USA were genetically modified to resist glyphosate.

However, the weeds have now built resistance to the weed killer, much like antibiotic resistance has developed in bacteria. The crops are now at a disadvantage, because the resources they need to use to resist glyphosate mean that they are less productive than the un-engineered versions that don't have to waste energy on this defense. Meanwhile, the new superweeds are now very difficult to kill. This is a classic case of the failure of the scientific method, ignoring the possibility of resistance developing in the weeds. Similar resistance has developed in crops genetically modified to kill insect pests, where the animal pests targeted, such as the pink bollworm, a moth species whose larvae attack cotton, are now resistant, and feasting unhindered on the cotton crops of India. Western corn rootworm has now become resistant to genetically modified maize that produces the same toxin.

Furthermore, systems are dynamic, as are the components that make them. Things are constantly changing in space and time, meaning that we cannot predict what can happen from past experience. This is the so-called black box scenario. We have seen so many interventions by humans, some of which we have discussed already, that have not worked out the way we hoped. Systems present challenges to our very way of thinking, where surprising, adventitious outcomes emerge. Yet we should always expect the unexpected. And we need to cope with it, because that is the way it is.

The third philosophical problem is that we are very bad at listening to feedback. Without listening to feedback, we cannot hope to re-integrate into the Earth system. Emergent systems work because of connectivity. And connectivity results in responsiveness. A dynamic system is constantly changing, and each component in it must be prepared to change too. In order to appropriately respond within the context of the system, we need to be listening to the conversations around. The black box is not a vacuum. It is made up of a myriad of connections, a bit like the internet of things, feeding information throughout the network.

Organisms either respond to the feedback and contribute to the overall system, or they fail to and risk excision. For most of our time on Earth, we were in the former category, but more recently we have been in the latter camp, living in our own little bubble. We need to re-connect. However, if we don't respect the other members of the conversation, we are less likely to listen and respond. We are solidly and stubbornly focused on ourselves, optimizing everything for our benefit.

The fourth philosophical problem is that we do not grasp the importance of sub-optimality. We have worked increasingly to optimize the Earth system for our own benefit. Of course, this is because we think that the planet is there for our benefit and ours alone. Humanism represents a religion of self-worship, where we are the gods and our wisdom and technology are to be focused on progressing towards a human utopia, at any cost.

We have converted ecosystems into industrial agroecosystems, poured fertilizers on fields to forcefully maximize productivity, damaged the waterways and soils in our pursuit of food and poisoned the land, sea and air in our quest for luxurious excess. It's all about us. Indeed, individualism is a founding principle of the Western neo-liberal tradition. Most of us don't even consider societal needs and responsibilities. It is all about me, me, me.

Even worse, we judge each other on how much ecological damage we do, not directly, but in terms of who owns the most damaging things. The bigger the better, as we strive to keep up with the Jones. "*My carbon footprint is bigger than yours*" is the unwritten mantra. This kind of link between ecological damage and status is difficult to shift. The rat-race, as it is called, celebrates wealth as success. Human wellbeing is measured in terms of GDP. How can you be happy if you are poor? Waylon Jennings bemoaned the pursuit of happiness through wealth in his song *Luckenback, Texas*, when he wrote:

"So baby, let's sell your diamond ring
Buy some boots and faded jeans and go away
This coat and tie is choking me
In your high society you cry all day
We've been so busy keepin' up with the Jones
Four car garage and we're still building on
Maybe it's time we got back to the basics of love."

Fundamentally, we need to sub-optimize, recognizing that for the Earth system to work, no single component can have the perfect life. We need to forget where we buried all of our nuts, dump the squirrel app, leave some grain in the field when we harvest our crops, eat the ugly vegetables rather than throwing them away and eat local, seasonal food, rather than expecting strawberries all year round. We need to put a jumper on rather than turning the heating up, and to spend some time joining a river-cleaning crew rather than playing that round of golf.

We don't require all of the luxuries that we surround ourselves in. In order to achieve an equitable society, some of us will need to do with less. Because an equitable society must be sub-optimal too. The functional society must function within a functioning Earth system, and so we all need to embrace a lifestyle with a reduced ecological footprint. This is what sub-optimality means.

The fifth philosophical problem, within the domain of moral philosophy, relates to how we think about the issue and the solution space. Motives are important. In a court case, the motive can greatly affect how we are judged and what punishment we receive. Imagine the situation where a man shoots an innocent dog. If it is shown that he deliberately killed this dog knowing it was innocent, he should be heavily judged. But what if a dog of the same breed, sex and stature had just killed the man's young child? He shot the dog because he was sure it was the one that had killed the young child, and he feared it would try to kill another child.

Surely the judgement upon this man would be totally different, because his motive was different. He wasn't a psychotic dog killer, but a loving father who was sure that this dog was incredibly dangerous and could pose a threat to others. Of course, he should have left it to professional dog wardens. He was wrong to have done what he did. But his motives were good.

When it comes to our planet, we are constantly sold a selfish, human-centric motive. If we don't save the planet, we are all going to suffer. If we destroy the rainforest, a rare medicinal plant could go extinct whose chemistry included a molecule that could cure cancer (the plot of *Medicine Man*, the 1992 movie starring Sean Connery as the misogynist, eccentric scientist and Lorraine Bracco as the pharmacologist). The motive that is messaged is to save the plant and the planet for our sakes.

These are the wrong foundations upon which to build a transformation. They hark back to the disastrous Enlightenment philosophy so engrained in the Western world. Nature is still being viewed as a source for us to build our prosperity and futures upon, not as an entity with sovereign rights, due our respect for what it is. Also, it encourages us to pick and choose what we save. We even see this in the language we use, such as 'ecosystem services'.

The cute, fluffy, useful creatures, rather than the ugly, slimy things are the focus of our attention. Even some of the leading conservation groups, such as the World Wildlife Fund (WWF) with its giant panda, have cute creatures on their logos. The useful and the beautiful attract our support, but this is underpinned by the pervading motive of saving what we need and like. This attitude is fundamentally flawed. Unless we address the elephant in the room (another favoured creature), our motives will be wrong.

We need to re-engage and re-integrate with the Earth system because it has the right to exist. Our rights are part of its rights, but only a part of them. We need to become naturalists, not humanists. We need to embrace the Earth system as an entity, not as a resource that we should save in case it might benefit us. That plant with the cancer-curing potential should be saved as part of the greater whole and for its contribution to the system. Also, we should never pick and choose.

Removing our interference is the key, not increasing it. Re-introducing some components of the system but not others is a strategy so flawed. If you wanted to re-introduce the beaver to Scotland, you also needed to introduce its predator, either the lynx, wolf or bear. You can't have one without the other. So, if you don't want bears wandering around the countryside, then you can't have beavers either. But really, you shouldn't be practicing ecological engineering anyway. Instead, you need to allow Nature to carry out its own restoration, and alter your behaviour enough to facilitate this. Rely on the feedback, once you've learned to listen, and embrace emergence, then let it be.

It is clear that these five philosophical barriers are significant but surmountable. Identifying them is half the battle. Awareness and accountability are another forty nine percent of it. And the last bit is down to overcoming lethargy, and waking up. Once we address the philosophical, ethical and moral aspects of this, it just takes a transformation of our thinking. Of course, there are structural barriers that are not of our making, and we consider these next.

IX.2. Structural barriers to change

Structural barriers are easy to identify, but often difficult to surmount. Government intervention is one way to overcome them, but many modern governments are neo-liberal in their political philosophy, and thus believe that state spending should be reduced, not increased, returning traditional public sector areas such as health, welfare and education to the private sector, and allowing market forces to underpin the shape of these areas. People may be aware of the issues and willing to act, but structural barriers (real or perceived) may prevent them from doing so.

Some examples of structural barriers include a lack of public transport options particularly in rural areas. The supplier of our electricity may not use enough sustainable energy production. Our shops may not offer alternative, ecologically-friendly products. You may be a single parent, working long hours in a low paid job while caring for an elderly parent, and so you have to use disposable nappies because you do not have the time to wash cotton nappies. These are real-world issues.

Other barriers can arise from oil prices dropping, as they did in 2015, resulting in alternative sources of energy becoming relatively much more expensive, meaning that they struggled to establish themselves in the free market of the globalized economy. Subsidies needed for alternative energy sources to remain even slightly competitive had to increase massively, but this was not possible due to financial restrictions arising from the great recession of 2008/9 and so much of this sector collapsed (Skene and Murray, 2017). Some would say that the oil-rich nations in OPEC flooded the market with oil, reducing the price per barrel, in order to stop alternative energy companies from threatening their control of the energy market (Foster et al., 2017).

However other damaging side effects also accompanied this oil price slump. Venezuela, heavily reliant on oil sales, had its economy wrecked, leading to ongoing and devastating political instability, starvation and rioting (Elliot, 2015). Thus, barriers can come from deliberate international politics and economic subterfuge and can have consequences far beyond their immediate targets. Wealth is power, and the protection of wealth is enough to go to war on, either economically or with an army if necessary. If green energy is seen as a threat, then an assault should be expected. We don't envisage OPEC countries blowing up wind turbines and solar panels, but reducing the price of oil can be just as effective.

One of the most challenging structural barriers is the reliance of the modern global economy on growth. Two quarters (six months) of negative growth are defined as a recession, while two years of negative growth earns the tag of a depression. Lord Stern wrote that *"It is neither economically necessary nor ethically responsible to stop or drastically slow economic growth to manage climate change. Not only would it be analytically unsound, it would also pose severe ethical difficulties and be so politically destructive as to fail as policy"* (Stern, 2009).

The 2008 crisis severely impacted on important spending plans of governments, and thus on society, and the shock waves are still being felt today, with public services in many countries struggling under the policy of austerity, even a decade later. This has, undoubtedly, contributed towards the rise in populist movements. In Europe, the treatment of Greece by the European Central Bank was viewed by many as humiliating, supremacist and degrading.

Leading experts are of the opinion that degrowth is an essential step to reduce ecological damage and, certainly, indicators show that the 2008 event led to ecological recovery briefly, with anthropogenic carbon dioxide production declining. Professor Tim Jackson writes that: *"The myth of growth has failed us. It has failed the two billion people who still live on less than $2 a day. It has failed the fragile ecological systems on which we depend for survival. It has failed, spectacularly, in its own terms, to provide economic stability and secure people's livelihoods"* (Jackson, 2009). Jim Yong Kim, president of the World Bank from 2012 to 2019, wrote that *"The quest for growth in GDP and corporate profits has in fact worsened the lives of millions of women and men"* (Millen et al., 2000).

There is an urgent need for GDP to include environmental damage (Skene and Murray, 2017). This is much like replacing the carbon footprint with the ecological footprint. We need proper accounting and unless we do this, we are living an economic lie. Thus, this barrier of a dependence on economic growth relies on incomplete accounting. By balancing the books, we will make economics accountable for the damage it does, both to the environment and to society, and then, as we decrease gross growth, the reduction in environmental damage will appear as a positive yield, incentivising us to continue with ecologically responsible economics.

This isn't some kind of trick. It is called good old-fashioned truth. Imagine a company that only reported what it made and not what it spent. Modern economics is very much like this hypothetical company. Missing completely from the leger is the vast cost of environmental and social damage. Only by seeing the full picture can we really judge how successful the current economic model is.

A final structural barrier is the silo effect. Because we are dealing with the complexities of the Earth system, we need system-based thinking. Trade-offs require a flow of knowledge across the full spectrum of all of the sectors involved. You can't just start fixing one problem, without considering all of the other problems. As we mentioned, focusing on the carbon cost of energy production can lead to us ignoring the water cost or the land cost or the extraction cost of rare earth metals.

Silos exist everywhere, from government departments to universities, and from local councils to international organizations. Often competing for funds, cards are kept close to the chest. Instead we need open doors and windows onto the wider world. The bacteria are such a success because they share DNA between themselves. We need to follow this example, and rip down the departmental walls. What I refer to as *'Open plan thinking'* is the only way to solve the complex problems facing us.

Yet even by resolving philosophical barriers and removing structural barriers, there still remains one other set of obstacles, preventing us from acting on what we know we need to address. These are the psychological barriers. Robert Gifford (2011) has identified a number of these, calling them the seven dragons of inaction.

IX.3. The seven dragons: Psychological barriers

Robert Gifford (2011), the Canadian psychologist, wrote a paper entitled *The dragons of inaction: psychological barriers that limit climate change mitigation and adaptation*, published in the *American Psychologist*. Gifford explains why there is often a gap between attitude and behaviour. While his focus was on climate destabilization, these dragons apply, and should be applied, equally to the other issues relating to the human ecological footprint. Let's work our way through these barriers.

The first barrier is limited cognition. By this Gifford means that humans are not as rational as they think they are. Their decision-making is clouded and impacted by a range of things. He refers to our ancient brains, wherein we are consumed with the immediate world, in terms of survival, because, originally, that was what our priorities circled around. Thus, we have problems thinking ahead. It's not impossible, but we struggle with more distant issues.

We are also, often, inadequately informed about the causes, consequences and necessary remediation required to address these issues. Deliberate mis-information can also muddy the waters (Hoggan, 2009). While ancient brains delay us from thinking ahead, we have lost much of our ecological intelligence that used to connect us to our environment, and thus we struggle to hear the natural feedback.

The other problem relating to limited cognition is that many of the products we see on the shelves are made in other countries, and so the ecological impact is exported, meaning that we are not aware of it. This again removes the urgency into the distance, in spatial rather than temporal terms. Our ancient brains will ignore what is temporally or spatially distant.

This is why the ecological footprint barcode that we have discussed earlier, using AI, is so essential. It makes things transparently obvious that we have been able to conveniently ignore by externalizing them. Another cognitive issue arises, perhaps surprisingly, from our optimism. While optimism is a positive human condition, helping us to cope with difficulties, it can be delusional. The Tom Robinson Band famously sang in the song *I'm alright Jack*, from their 1978 album, *Power in the Darkness:*

"Don't you worry, I'm alright, Jack
We've never had it so good
There's plenty of grouse at the country house
We're eating as we should
Hugh's at Sandhurst, everything's safe
With Perkins running the farm
Half a dozen shotguns in the Land Rover
Ready for the call to arms."

A final flaw in our cognitive approach is that of fatalism, where we feel that the whole thing is so big that whatever we do will not be enough. Certainly, our ancient brains, lacking an understanding of how the planet works, felt in awe of Nature. Indeed, many early civilizations worshiped Nature as a god or gods. Negative messaging, where we constantly hear how we are all doomed, plays to this fatalistic feeling. And this is a huge danger. We need positive messaging, showing us that there is a way back, a reason for action, a chance to make a difference.

Fatalism feeds off negative environmentalism. We need to cut off this food supply, starve it into submission and use the facts that we have to show that there is a path back. Belief in some form of religious deity today also plays into the fatalistic attitude of millions, who feel their destiny is pre-determined. These ideas all find fertile ground in an ancient cognition.

The second dragon is ideology. When Gifford refers to ideology, he includes the philosophical barriers we have already discussed, such as the dominant worldview of the West and North, neo-liberal free market economics. This worldview is firmly rooted in the Enlightenment, believing that technology and reason will allow us to solve all of the problems, as we march, uninterrupted, towards Utopia, on our path of progress. We have shown in this book that technology can facilitate our journey, but that the solutions are not technological in themselves.

The Earth system is too complicated to be mimicked by technology. It doesn't mean we reject technology, as so many strong sustainability advocates urge, but rather that we use it within the context of re-integrating with the Earth system. This is not some middle road between weak and strong advocates. Rather it is a new vision that embraces AI is an essential component. We do not have to return to the caves of our ancestors, nor the parliament of St Kilda. We have the tools at our disposal to assist in this great work, and they offer what we need: real time feedback. So why not use them?

The third dragon is our tendency to compare ourselves to others. The social nature of humans can be a blessing or a curse. We've seen how groups of AI robots exhibited tribal tendencies when placed together, splitting into factions (Whitaker et al., 2018). Social bonding is a complex issue, and pack politics is found throughout the living world.

Of course, social cohesion relies on resonance within the group. As we have already noted, Adam Smith emphasised the importance of such resonance as the basis for our ethical values and for the invisible hand that tempers our behaviour. However, in humans, this can form an important barrier to change. We are easily influenced by those around us. If our neighbour drives a large car and has his air conditioner on constantly with the patio doors wide open, then why shouldn't we?

Gifford also points to the irritation of perceived free riders in a community. What's the point in making the effort to live more sustainably when the guy down the road doesn't? Inequity also leads to a decline in cooperation and trust within communities if the inequity is visible and noted (De Cremer et al., 2001). One shard of light is that this barrier can be turned into a force for good. If the resonance of society can be brought into tune with the environment, then there can be positive feedback, wherein an environmentally attuned society will draw others towards this attitude. We shall examine this in greater detail in the final chapters on transition.

Fourth comes fixed costs. Consistency is generally a positive quality. There are so many players around us, trying to pull us in every direction. Thus, there is a natural, protective resistance to change. It is a good thing. If we changed our view every few hours, we would have no foundation on which to build anything. The changes needed to alter our destructive dive towards a crash landing are immense. We have invested, physically and metaphysically, in the worldview that has led us here. Gifford gives the example of a car that you have recently purchased. To leave it in the driveway, taking a bus or a bicycle instead, makes poor sense in terms of the money you have paid for the vehicle, road tax and insurance.

Old habits die hard, as the saying goes, and sunk costs contribute to what has been termed *behavioural momentum*, where behaviour persists in spite of environmental cues to change. It is a form of active inertia, a runaway train on a slope, where we just keep barrelling onward, in spite of the warnings of a bend (see Mace et al., 1988). Behavioural momentum needs a firm hand to change, such as legislation or incentivization. Free park-and-ride facilities, laws banning cars on certain days, cheap and regular public transport and car share bonuses can all help. You can teach an old dog new tricks with persistent, consistent messaging.

Another issue relating to our investments is that we face many challenges in life. This is called *intersectionality*. We have many, sometimes conflicting values relating to each of these challenges, and there can be a tension between attempts to resolve each of them. The required transformation of our lives, attitudes and actions in order to re-integrate into the Earth system cuts across most or all of these other issues, and thus requires a vast amount of adjustment.

Sunk costs can be identified with each of these issues, and we are probably sacrificing a lot in order to balance everything already. Thus, the thought of such massive change is unlikely to be a comfortable *chaise longue* on which to recline, but, rather, a *lit de clous*, or bed of nails. However, it is important to remember that the bed of nails has for a long time been used in acupuncture as a health treatment. The most dangerous moment is getting on or off such a bed, and, in many ways, this is analogous to the challenges facing us as we envisage the difficulties that such a necessary but impacting transition will require. There is no gain without pain.

The surroundings that we have invested in also carry with them status within a capitalist worldview, where success and wealth are synonymous with how we think that we are perceived. In *The Good Life*, a British television series in the 1970s, two couples live next door to each other. One couple decide to escape the rat race and to life sustainably, turning their garden into a farm and rearing animals in suburbia, including a rooster called Lenin, much to the horror of their neighbours who are heavily invested in the symbols of wealth. The series explores the tensions of these two worldviews and painfully exposes many of the issues that confront us today as we face the necessary changes and accompanying stereotypes that a more environmentally accountable lifestyle conjure up in our minds.

Discredence is the fifth dragon and relates to a lack of trust in a particular viewpoint. In this context, Gifford is referring to society's perception of environmental warnings. Discredence, or a lack of credibility, can stem from a number of sources. Firstly, there is a huge responsibility on the side of scientists and governments to adequately explain the issues at stake, in a non-threatening and transparent way. Scientists are under pressure to secure funding continuously, not only for their own promotion chances, but for the university or institute in which they work. This can lead to exaggerated claims, dumbing down and arguments with other teams with whom they compete, becoming very public and very messy. This is confusing for people.

Governments are focused on wielding and maintaining power in order to pursue their particular political agendas, while keeping the electorate on side. Thus, it is rare that a government will make a decision that is both difficult and unpopular. This can put the government and scientists on different trajectories, further confusing people. This situation is becoming exacerbated by an increasing distrust of mainstream politics. Across the world, populism is on the rise, as is extremism. The scepticism tends to target scientists as well as politicians, further weakening the argument for change. People can also adopt an attitude of denial, believing that there is no threat or that we are not responsible for the problem.

The sixth dragon relates to *perceived risks*. What could the implications be if you were to climb on board the transformation train? Could it damage you financially? Could you lose friends and be judged as a fruit loop? Could your reputation and social standing take a hit? We are generally risk-averse, and in many situations, this is a good thing. But fears over what might happen if we make the necessary changes can overwhelm us and prevent us from acting. Hamlet, the Prince of Denmark, bemoans the impact of such fears on his ability to take action, in Shakespeare's play, The Tragedy of Hamlet, Prince of Denmark:

"Thus conscience doth make cowards of us all,
And thus the native hue of resolution
Is sicklied o'er, with the pale cast of thought,
And enterprises of great pitch and moment,
With this regard their currents turn awry,
And lose the name of action" (Act III Scene I)

Gifford's final dragon is that of *limited behaviour*, or minimal response. People tend to do the least they can get away with. This has been also referred to as the low-cost hypothesis, where we take minimal action, assuaging our feelings of duty, but not enough to actually have a significant impact (Diekmann and Preisendörfer, 1998). People choose the pro-environmental behaviours that demand the least cost (Kollmuss and Agyeman, 2002). This can be an outcome of the intersectionality problem, as already discussed, where we are forced to balance our responses to many different problems.

Just like systems, we can't optimise for any of these issues, and must act sub-optimally. *"Well I can't do everything, can I?"* is a frequent response of a harried parent, trying to just get through the day, when asked about sustainability issues. This is the real world after all, not some idyll of pastoral perfection, where lions lie down with lambs. Lions eat lambs every time in the real world. There can also be a rebound effect, where, for example, if someone buys a more fuel-efficient vehicle, they end up driving more miles and thus releasing the same amount of pollution as when they drove less in their less efficient vehicle.

IX.4. How AI can help overcome these barriers

The five philosophical barriers mostly circle around the fact that, in the West, we have been trained to think in a reductionist, empirical way. This stems back to school education. It is imperative that our curricula embrace systems theory from an early stage. In so doing the next generation will be much better equipped to deal with the so-called *wicked problems* that are really just system characteristics. For the rest of us, we need re-educating, through evening classes, multimedia presentations and compulsory work training. This may sound excessive, but it is most certainly not. We need to understand what a system is, how the Earth system works, and why we need to act urgently.

Scientists and governments need to understand this too. Of course, most ecologists, physicists and non-symbolic programmers acknowledge the centrality of systems theory already. But it is essential that everyone grasps the key characteristics and properties of systems and, in particular, the Earth system. AI is perfectly suited for the role of supporting these education programmes, particularly in terms of providing analysis of the state of the Earth system, and in terms of delivering system theory to the human population of the planet.

AI can also provide the evidence we need to demonstrate how necessary the changes are. Of course, indigenous races already fully understand what being part of the Earth system means. These people from around the world can play central roles in educating us, giving us an insight into their ecological intelligence, and what it means to be integrated within the greater whole. Indigenous rights and protection must now become a priority.

Structural barriers generally need state or private sector intervention. Improving public transport, ensuring public procurement is ecologically sound, providing support and financial incentives to families and communities to encourage action and putting in place a sensible regulatory framework are all important in order to overcome structural barriers.

Eradicating fuel poverty, ensuring equitability in terms of access to infrastructure needed for a sustainable lifestyle and developing agricultural, energy sector and industrial standards that address the significant issues in these areas will best be done through government intervention.

It is not all about government, but there are responsibilities that only government can accomplish. By removing these structural barriers, people are then able to move forward. AI has a central role here, both in terms of ecological accounting (the measure of the ecological cost/benefit of particular economic activities), information transfer and providing feedback to government in terms of identifying areas needing a more focused approach.

Slaying the seven psychological dragons of inactivity, as identified by Gifford, really circles around five swords. First, reliable information on the state of the planet is needed, preferably delivered by people with no ulterior motives. This is more easily said than done. Transparency is key, and a new, independent, publicly-accountable organization with no historical issues, either perceived or real, is required. Experts in communication should design clear, unpatronizing analysis of the data pouring in from the internet of things, from species diversity to habitat changes and from remote sensing through to mobile device sound recordings.

Proper, transparent explanations of how data is collected, why and for what purpose, need to form part of the strategy, so that people are empowered to interpret the evidence unequivocally, without being told what it means. Raw data files should also be made available so that anyone who wishes can analyze it independently. All of this data must be freely available, as it is the property of the planet. This avoids conspiracy theories from developing.

The second weapon with which to slay the dragons is education. Information is important, but an understanding is also essential. This will help address the limited cognition issues, particularly in terms of temporal and special aspects of the crisis. Outdoor education is particularly important here, and needs significant expansion. Again, AI is perfectly equipped to provide interactive education, providing the skills to interpret the feedback from the Earth system. Augmented/mixed reality can play a significant role too, allowing us to explore repercussions of our actions or to experiment with innovative concepts in a safe space.

The third sword is a functioning society. There are many theories relating to how best to achieve such a thing, but clearly, in terms of the dragons such as comparisons with others and perceived risks, trust and collective action are important. Furthermore, meso-level feedback, as discussed earlier, will help address many of the issues here. Ultimately, strong community leaders are important, and training can be provided for people, all designed to strengthen societal connectivity. Outdoor education and activity will strengthen the environment-society bond, building functionality and motivation. Society has adjusted to significant transitions in the past, and will do again.

The fourth sword is incentivization. In order to reduce cheating, mistrust and disengagement, government incentives relating to rates and tax reductions at the community level can play a significant role. This can help offset sunk costs but also acts as a motivation to work together. Grants for assistance should be easily applied for and transparently assessed. Just and equal treatment is essential if we are to avoid perceived inequality.

The final sword is positive messaging. This is essential in order to avoid feelings of fatalism and to address behavioural momentum. Only if we feel we can and do make a difference will we want to act. Threat messaging will only strengthen such barriers. Painting apocalyptic pictures is more likely to drive people into their shells than out on the battlefield. Also, children and vulnerable adults may not be able to cope with such negativity, leading to fear and even psychosis (Clayton et al., 2017). The idea is not to hide the cruel truths of the devastation we have created, but rather to combine this understanding with explanations of how we can recover. Fundamentally, the Earth system has recovered from much worse in the past. It has the creative force within it to recover again. The only question relates to whether or not we want to be part of this recovery. If we do, then the path is there for us to re-integrate and celebrate.

SECTION X Transition

"Those in the System, would like us to share their belief that all the changes are not connected: they are simply anomalies, isolated symptoms to be treated or preferably ignored, before the all-powerful Western capitalist patriarchal model goes on to ever greater heights and grander ejaculations. Most are numb to it, caught in fear, denial or resistance.

But we, Burning Woman, know this process intimately. Amongst Burning Women and Men, there is a fierce, quiet knowing that these are both the death pangs of the old, and the birthing pangs of the new."

–Lucy H. Pearce, *Burning Woman* (Pearce, 2016)

Welcome to the final section. It has been quite some journey. We started with King Mu of Zhou, whose jealous rage and mistrust led him to order the destruction of the first conceptual AI. We've seen how all the tribal flaws of humanity such as sexism, racism and inequality have re-appeared in the new technological age, mere extrapolations of our failed state. We have encountered different ways of thinking, and came face-to-face with ecosystem intelligence, with its emergence, sub-optimality, non-linearity and real-time feedback.

We've explored the urgency of addressing the collapse of the Earth system on which we rely, and envisioned how AI can save the planet, by acting as the communicant, arbitrator and invisible hand across the three arenas. We learnt lessons from the Ogiek and St Kildans. Barriers to the necessary changes were identified and analyzed, and now, finally, we turn to the transition itself. Transitions are more common than you might think. Each of us goes through many of them in our lives. We look at the theory surrounding this challenging topic. We recognize that it is not a transition of the Earth system that is needed, but rather, a societal transition. We outline how this can happen, before ending the book with another character from ancient times: Midas, the King of Phrygia, with his donkey ears.

X.1. The nature of transition

All of us have undergone some form of personal transition in our own lives. We transition from junior school to high school, and maybe to university and then work. Adolescence is a significant transition in any life, as you move from childhood to adulthood. You might have moved house to a new area. You might begin a close relationship, marry or divorce. You might vote for a different political party than you normally would. Parenthood requires a huge transition. Retirement is a significant transition for many. You may sustain life-changing injuries. All of these transitions are personal, but some, if not all, impact on others. There is a before, a during and an after to most transitions. There are impacts upon your physical and metaphysical worlds.

Personal transitions are often beyond your control. Take the death of a close friend or relative. When this happens, it is devastating. You couldn't stop it, and are left with a different world than it was before that precious person left. This book is dedicated to two of my friends who died within six months of each other. I'm still reeling from the loss of them. My whole world is still transitioning, and may well continue to change as the dust settles very slowly on the events that took them from us. I spend much more time just staring into space than I used to. At other times the experience sharpens my decision-making, my moral judgements and my drive.

However, a societal transition is a completely different creature than is individual transition, even though it obviously impacts on individual members of the society. Societal transition is a form of system transition, an emergent quality, and this impacts hugely on how it can be 'managed', as we shall see shortly.

X.2. Studies in transition

Loorbach et al. (2007) define a transition as a transformational process "*in which existing structures, institutions, culture and practices are broken down and new ones are established*".

There is a significant literature on transition. Most transition theorists discuss the changes across our history in terms of changes from hunter gatherer to agrarian to industrial to technological. The Viennese School of sustainable transition, called *socio-metabolism,* is led by Professor Marina Fischer-Kowalski, director of the Institute of Social Ecology at the University of Alpen Adria in Austria. This school is tightly tied to the traditional ages of transition of humanity, using socio-metabolic profiling,while focusing on energy and resource use (metabolisms) over time.

Socio-metabolism is defined as the process by which societies establish and maintain their material input from and output to Nature and as the way in which they organize the exchange of matter and energy with their natural environment (Fisher-Kowalski and Haberl, 1997). The Viennese School interprets each age of humankind as a stable socio-metabolic regime (hunter-gatherer, agrarian, industrial and technological) with a rapid transition between each age. It then calls for another great transformation to a sustainable future, reducing the energy and resource use and altering society.

There are a number of problems with the underlying principles here. Firstly, the narrative flowing through our history is not that of four ages, but of one change, from hunter-gatherer to agrarian. With the onset of the Agrarian Age, social structure changed, from an equal, communal and shared governance as seen in the Ogiek people and the St Kildans, to a hierarchical, territorial and unequal society.

Since the great transformation from hunter-gatherer to Agrarian, there have been gradual changes in efficiency and intensity of production of essential comestibles, along with an increasing provision of luxury goods for the few. New technologies have been drip-fed into the story through time. It is unfair to claim that technological advances only impacted our history in the last two hundred years alone. Think of the irrigation channels of the Babylonians, the aqueducts of the Romans, the domestication of horses for transport and the production of iron, tin and bronze.

However, inequality has increased, and the poor are just as poor. Slavery and people-trafficking continue unabated, and war, disease and famine are still with us. The rich have indeed got richer, benefitting from new developments. For example, the richest people on the planet were industrialists up to recently, but now they are technologists.

Yet there has been no real transformation across our history since the Agrarian Age was established, just a continuation and intensification, with a scattering of dictators and disasters along the way. In many ways the move to a settled lifestyle can be seen within the context of the end of the last ice age. Following the hardships imposed by climatic change, humans embraced the more amenable conditions of the interglacial by transitioning into agrarians.

There has really only been one transition so far, from hunter-gatherer to settler. This change occurred perhaps twelve thousand years ago, where humans settled in built environments, began farming and specialized into different roles such as mason, soldier, tinker, tailor, candlestick maker and ruling elite. The rest of our history is really a story of increasing productivity and efficiency, trading, mechanization and inequality.

Of course, there have been some big changes along the way. The movement of populations from rural to urban settings during the nineteenth century and simultaneous industrialization of food production defined the recent history of the Western World. However, these are merely the continuation of a long-term trend.

Another point is that none of the changes through time were planned at a global level. Neither did each age completely displace the previous regime. All four ages occur on the planet at present. None of the ages saved the planet from the previous regime's excesses and problems, but, rather, exacerbated them. None of the ages used extrapolated data. None of the ages had to deal with habitat destruction, pollution or climate destabilization as existential issues.

Thus, the world we live in today is different than ever in our history, the challenges more marked and the issues more urgent. Transformation is required, but it is the Earth system that will impose the transformation on us unless we undertake change ourselves. The driver for change here is not technological innovation, but rather existential threat. This is a completely different scenario than has ever been faced by us. Technological revolution can facilitate the necessary changes, but a global, social response is needed. The socio-metabolic theory of the Viennese School correctly pinpoints our need to reduce energy flux through the environment and reduce resource use, but it fails to grasp the prime directive, re-integration within the Earth system.

Also, there is a lack of recognition of the reality of a heterogeneous human race, and the need for a heterogeneous response. Although we all live on the one planet and are all suffering the consequences of the ecological devastation we have mastered, not everyone is at the same point in terms of the so-called ages of humanity. Transition cannot be a global single track, where everyone follows the same path. The mantra of sustainable development, as preached by Northern, Western nations fails to recognize the Earth system as the solution space, instead pursuing the disastrous policies of the Enlightenment.

Many indigenous people, often hunter-gatherers, are already situated within the Earth system and have small, appropriate, ecological footprints. These people don't really need to change anything, and they certainly do not require the neo-colonialist rubric being forced on them for their conversion to a 'progressive' Western society. They are fully ecologically developed already. Agrarian civilizations require different approaches, as do industrial and technological societies.

Socio-metabolism focuses on the use of resources and energy, but fails to recognize that any transition must be systems-based, and, specifically, centred within the Earth system. Thus, the focus must be on the key characteristics of the system, namely emergence, non-linearity, sub-optimality and feedback. This has to be the focus. Any change in societal structure and function must be oriented towards facilitating our re-integration into the Earth system, and not on historical observations on previous socio-metabolic regimes. As we have already shown, these regimes were really just a continuous development and realization of the settler regime that followed our ongoing transition from hunter-gatherer. Rather than trying to structurally return to the hunter-gatherer, we need to re-discover the social function of our origins, focused on equality, communalism, shared governance and ecological continuity.

One question that comes up when I have discussed the use of AI in the transition to re-integration is that people in many 'developing' nations do not have access to advanced computer technologies such as the internet of things and AI. However, this is not a problem because the further away from re-integration a nation is, the more dependent on technology it is, and the more available that technology is. Nations with good access to these technologies are also the ones with the greatest need of re-establishing feedback with the Earth system.

Thus, the use of AI as a catalyst for change, while less available for indigenous and agrarian societies, is less necessary in these communities, as they are already well integrated. As AI is seen as enabling those of us who have an ecological disconnect (i.e. the so-called 'developed' world), it is not necessary for it to be globally available. Provided that it is available to those that need it most, then AI can play its role. And this is in fact the case.

So, when we think about the transition needed to prolong our membership of the Earth system, we are discussing a very different kind of change. Rather than continuing on the path of the last twelve thousand years, we need to shift completely. Most of the transition theory doesn't really apply, because we are dealing with a system that, although ill, already exists and, like all systems, is emergent.

Why is this so important? Well the whole point about an emergent system is that you cannot build it, nor can you say where it will go, what path it will follow or how it will get there. There is no point in setting targets or imagining a final destination, because it is a black box, full of complexity and surprise. This is not like laying down a train track from A to B and driving the train to some planned station. There is no track, or at least no single track. It's more like a hovercraft whose steering appears out of our control. Welcome to the world of systems.

X.3. Why societal change is key

When we look at the Earth system through history, although massive events such as snowball Earth and comets hitting the planet have shaken things up, the actual system has continued to do the same things. Thus, the system has never transformed, only the structural outcomes. Functionally, energy still flows through food webs and complexity build over time, reaching a dynamic equilibrium. Species come and species go, but the functionality remains the same. Hence it is not a transition of the Earth system that is needed here. Rather it is a transition of human society. And it is not a transition between forms and structures, because a system is emergent. You can't insert forms into the system. They will most likely be spat back out or cause disruption. It's about function, and functional control lies within the system, not with us.

Before considering how we should approach this, let's reflect on three things that have to happen. The first step involves the reduction of our disturbing behaviour, by restoring appropriate flows and fluxes. This means taking immediate action on fertilizer and pesticide misuse, toxins from mineral extraction, CFC production, greenhouse gas emissions, habitat destruction and soil abuse. There is no time to waste on any of these. Our role is not to add chemicals to the oceans to draw down carbon, nor to increase the reflectivity of the planet (its albedo) to reduce incoming solar radiation. Rather by alleviating the stress upon it, we will allow the Earth system to repair itself. It knows how to do this. The three key areas for action are agriculture, energy production and supply chains. Consumers, government, producers and designers share equal responsibility in all three aspects.

The second step involves the need to stop conservation efforts, instead trusting the Earth system to find the best way forward. Re-introducing beavers, for example, without an appropriate predator, is a really dumb thing to do. This causes a complete imbalance in the food web, and will have disastrous consequences for the ecosystems into which these herbivores are released. Captive breeding is ill-informed and disastrous in terms of upsetting the gene pool and because once re-introduced, individuals bred in captivity have no idea how to survive in the wild.

Planting trees in a field is not the way to produce a forest. Forests build from bare ground, and take about a century to form. This is because the forest ecosystem isn't really about the trees, but it relates to the soil, the interactions and the developmental processes. To make a forest, buy some ground, leave it alone and let Nature do the building.

Genetically engineering coral to withstand increased temperatures again is a terrible idea, forever changing species that have been around for millions of years and altering their genetic makeup. Instead, corals can be moved to cooler waters and then restored once the first step is well under way, and the feedback from the Earth system tells us it is time. Provided we have carried out the first step, Nature will be able to replace species that have been lost, reforest the planet, restore appropriate hydrological cycles (with the help of the forest it has just created) and absorb carbon dioxide (again, ably assisted by the natural forests now thriving).

The restoration of species is called diversification. It lies at the core of biological evolution and is a diffusive process. The term *evolution* comes from the Latin word, *evolvere,* meaning to unroll a scroll. The story of life is like this. With time the scroll is unrolled and the story is revealed. Just as Darwinian socialism has been rejected, so too Darwinian evolution is now seriously questioned as an explanation of how the process of diversification and speciation occur. And the reason is extremely important.

If we examine the diversification of species in the fossil record, we see that when there is a mass extinction, the number of species falls dramatically, as you would expect. However, following this, there is an increasingly rapid diversification of species up to a point, after which speciation slows to a net rate of zero. Of course, during this flatlining, species die and are replaced, but the overall balance is maintained. It is what we call a dynamic equilibrium.

The flat line is when all the niches are filled. It's like a shopping street, where all the buildings are filled with different businesses: a news agent, a grocery store, a greengrocer, a haberdasher and a hat maker. Then the hat maker dies and his business shuts. A new business opens in the same building, maybe a candlestick maker. When the street is full, there is competition and selection. Businesses that can't compete close and are replaced.

However, every so often a massive economic depression hits and most of the shops shut. This is the equivalent of a mass extinction. And, like a mass extinction, after it, there is no competition or selection, and diversification is at its highest. In other words, evolution occurs in the empty market places, not in the crowded back alleys (Skene, 2009). And so, evolution thrives on opportunity. New species diffuse into the free space. This is why Nature can recover from devastating events. Functionality is restored, and new forms appear. It's how the Earth system works.

We really shouldn't interfere. Nor do we need to. Furthermore, natural selection is a consequence of niches filling up, leading to increased competition. Natural selection is not the basis of diversification, but the consequence of it. And natural selection supresses diversification rather than promoting it. Opportunity is the basis of diversification. Nature gets creative when it needs to, and when freed from the constraints of competition and selection.

The third step is to recognize that it is society that must transition, not Nature, and this includes the instruments of society. Grin et al. (2010) define a sustainable transition as a "*radical transformation towards a sustainable society as a response to a number of persistent problems confronting contemporary modern societies*". Here we see that society is the home of transition. De Haan and Rotmans (2007) point out that a transition is a long-term process of non-linear social change, taking one or more generations. And so societal transition needs to start immediately. But how do we do this?

X.4. How to manage societal change

There are two approaches to societal transition, neither of which are exclusive. The first is the top-down model. This model involves institutions imprinting solutions upon society. Here governments take control of the process, and enforce change through laws and policies. Certainly, in China, where the government has immense influence, the implementation of a circular economy through a series of five-year plans has had significant impact (Murray et al., 2017). However, there are downsides, involving an imbalance of power, exogenous drivers, inequality, disempowerment, the reduction of community resilience and a lack of consensus, being centrist in nature.

The second approach is the civil society model that focuses on empowering the community. It entails social justice, builds social capital and is rooted in traditional knowledge. It has endogenous drivers, bringing community resilience and, preferably, is founded in human-environment relations, being localist in nature.

Barkin (2010) identifies five key strengths of a civic society transition: autonomy, solidarity, self-sufficiency, productive diversification and sustainable resource management. Decisions being made by the people *in situ* are seen as being much more relevant and valid than those taken from a distant, centrist, political institution. A functional community also requires much change, particularly in Western societies where the trend has been towards individualism beneath the shade of the neo-liberal tree.

Community identity can be difficult to establish, particularly in urban centres with transient populations and temporary employment. Thus, the realization of community-led transition, where everyone is involved, may be challenging. Anyone who has been involved in local communities will know that there are often disagreements, cliques and over-dominant individuals. Not all of this is Machiavellian, Janus-faced trickery. Most people hold views that are well thought through, honest and reasonable. A functioning society should be able to accommodate diversity and use this as a strength.

I've chaired and been a part of many committees, some more successfully than others, and most of them community-based. The best ones have had a few things in common: a neutral, process-focused chair, a positive spirit, a clear reason for existing, a mixed set of life skills that are recognized and embraced by all of the members and a respect for each other. Esther Wojcicki developed the TRICK acronym for successful parenting, identifying five key aspects that are also highly applicable to successful community groups: trust, respect, independence, collaboration and kindness (Wojcicki, 2019). Civil society groups can deliver, but need a positive mindset embraced by all.

Societies do undergo dramatic changes at times. Podobnik et al. (2019) reflected that "*We expect that if economic issues prevail, a society partially becomes more prone to intolerant LW* [left-wing] *populism. In contrast, if ethnic issues start to dominate, a society starts to lean more on intolerant RW* [right-wing] *populism*".

While environmental or green politics is normally associated with left-wing populism, it can also appear in extreme right-wing writings. Take the following quote: "*We recognize that separating humanity from nature, from the whole of life, leads to humankind's own destruction and to the death of nations. Only through a re-integration of humanity into the whole of nature can our people be made stronger. That is the fundamental point of the biological tasks of our age. Humankind alone is no longer the focus of thought, but rather life as a whole... This striving toward connectedness with the totality of life, with nature itself, a nature into which we are born, this is the deepest meaning and the true essence of National Socialist thought.*" It all reads like a marvellously modern, green political agenda, with all the boxes appropriately ticked, until we get to the last line. It was written by Ernst Lehmann (1934), the Nazi botanist who initiated the Aryan Biology movement. Fascism embraced an element of green politics, mostly to engender pride in the 'Motherland'.

The problem with a political 'green' party is that it is a political party, and risks being engrossed in the dynamics of power and elections. Furthermore, by associating with the left, it may alienate those on the centre and centre-right. This is unhelpful when it comes to a societal transition. Also, grounded in left-wing populism, it must resist being swept away by the strong tides of Socialism, founded on Humanism, the very celebration of our separated identity. Instead the emphasis must be on ecological societies, and be neither left nor right wing in its politics.

The Earth system is neither socialist, conservative, Democratic nor Republican, and neither can be our approach to transition. Marrying our transition to a particular political philosophy, populist or otherwise, will only compromise the transition, because political philosophies have all contributed to the destruction of the planet, either through neo-liberal capitalism or socialist capitalism. Both the Soviet Union and the Western bloc were extremely disruptive to the Earth system, although approaching this disruption from two diametrically-opposed political philosophies.

Civic societies also face this challenge. The Lisbon Treaty of the EU, which came into force in 2009, also references localism, stating that "*The treaty creates a basis for a more decentralized and transparent approach to implementing EU policies to help ensure that decisions are taken as close as possible to the citizen*" (Lisbon Treaty, 2009). Localism has been embraced both by left- and right-wing political parties. In Britain, the Labour party set out what it called *'new localism'*, designed to combat civic disengagement and encourage community engagement. By 2006, this was renamed *'double devolution'*, reflecting the idea that governance would shift from centrist to local, and then from local to very local (Lyons, 2007), with an emphasis on community groups and citizen empowerment.

However, these policies were strongly attacked as representing backdoor centralization (Davies, 2008). Lyons himself, the author of the report underpinning double devolution, admitted "*No one should underestimate the sustained effort which will be required to achieve a real shift in the balance of influence between centre and locality. The history of the last 30 years is marked by a series of well-intentioned devolution initiatives, which have often evolved into subtle instruments of control*" (Lyons, 2007).

In 2010, a Conservative-dominated coalition replaced Labour, led by David Cameron, one of the architects of the Sustainable Development Goals. Representing the right wing of politics, he too embraced localism, in his *'Big Society'* policy. His government introduced the Decentralization and Localism Bill that became the Localism Act in 2011. This act emphasized the empowerment of communities through local referenda, vetoing council tax rises and a community right to bid, while following the doctrine of neo-liberalism, wherein government expenditure would decrease and taxes would reduce, facilitated by decentralization.

A similar move was happening across the big pond in the USA, where the Tea Party also presented a decentralizing agenda (Armey and Kibbe, 2010). David Walker, former managing director of the Audit Commission, gloomily summed up the adoption of localism by both wings of politics with the line "*We are all Localists now*" (Walker, 2002).

However, the increasingly recognized necessity of acting urgently to save a catastrophic nightmare being realized and met upon the human race gives civic society a new and vital role. The transition of society into a responsible, accountable collective is essential if we are to make the changes we wish to make. Top-down centrist approaches will not work, as can be seen in the increase in CFC gases in China in spite of strong legislation and enforcement of the five-year plan environmental policies.

Societies must engage with and embrace the Earth system, and hearts and minds must be in synchrony. This can best be done at the community level, but with government assistance. This must not be political, and should be informed as much by social science as by the natural sciences, since the core issue is societal function. This assistance should take at least eight forms:

1. Legislation and policy clearly providing support for civic engagement. This should include financial incentives, funded from a green tax on business, that reward communities who deliver reduced ecological footprints and environmental protection with lower rates, and reward individuals who participate in these civic groups with reduced income tax, or increased allowances and pensions;
2. Training in how to form, chair and work in groups;
3. Provision of infrastructure, such as small-scale hydropower projects on local streams to generate energy locally, the provision of meeting spaces and improved public transport;
4. Design and provision of AI facilities to provide essential feedback on the Earth system and on individual and community ecological footprints, as detailed earlier;
5. Reform of the education system, wherein schools have at least fifty percent of the curriculum dedicated to environmental issues and projects targeted at re-integration. All curricular subjects can be taught through an environmental lens, such as Geography, Economics, History, the Natural Sciences, Mathematics, English and Art. Increased teaching outdoors should also be an important part of the curriculum. Universities and colleges must have a separate strand running through their courses on environmental issues, focused on the degree programme being delivered. Education is essential if society is to have the necessary information needed in terms of what is at stake. Compulsory sessions within adult employment will also be needed, preferably as continuing professional development (CPD). Finally, an emphasis on systems thinking is urgently needed;
6. All businesses should be guided by government regulations to support and encourage community activities and to provide appropriate training. Business managers should be particularly directed in terms of carrying out supply chain inventories of ecological footprints associated with their products for public awareness. Significant green taxes on poor performance in terms of product footprint and employee support should be levied, in addition to deductions from dividends of shareholders;
7. Fuel poverty and other poverty-related barriers to sustainable living should be regulated against. Large ecological lifestyle footprints (ELFs) should be vilified in much the same way as drink-driving, smoking and not wearing a seatbelt have been successfully campaigned against in the past;

8. Communities should be involved in decision making and planning at regional and national levels in three ways (Mauser et al., 2013). *Co-design* integrates policy makers and citizens in a creative union. *Co-production* brings citizens into the heart of the research process. *Co-dissemination* gives citizens full access to findings, allowing then to channel their responses into further co-design. In all three areas, AI can play the crucial role of conduit and analyst.

With this support, and the micro-, meso- and macro-feedback detailed earlier, society can move forward, accountable, responsible and motivated to make the transition required, re-integrating into the Earth system and benefitting from the vast rewards of a healthy, functioning and inclusive planet. AI and the internet of things are ready to take on their roles. All we need to do is begin.

X.5. Requiem for the King of Phrygia

Ultimately, a functioning, eco-centric society can only be a positive thing in terms of social as well as environmental sustainability. The benefits are manifold, and priceless. You can't buy a focused, active, collective community. Any government worth its salt should realize this and work towards these eight goals. Any government that helps deliver such a rejuvenated society will also benefit on so many levels. Undoubtedly, social spending will decrease, as a functioning society will have less need for support, and this is likely to be replicated in areas such as health, social work and policing.

All of the barriers we have mentioned earlier can be overcome by a society working together, in tune with the Earth system and accountable for their impact on the planet. It is as simple and as difficult as that. Societal transition needs organization and support, but once working, it will build its own resilience. Emergent structures will result. We don't know what this will look like, but we do know that this will be a place of healthy environments and healthy societies, replete with equality, empowerment and cohesiveness. What's not to like? Artificial intelligence provides the feedback and the motivation like no other technology has ever done. It is a harbinger for good when used in this context, and can facilitate and empower the process and the people.

AI is ready and able to be that bridge across the troubled waters of disengagement, bonding attitudes and actions and straddling the gulf of apathy. The internet of things provides the opportunity for connectivity and allows us to listen to the very heartbeat of the Earth system. This far-reaching invisible hand, across environment, society and economy, will guide our decisions.

So now that we have a functioning society, what next? Re-awakened citizens, informed by the internet of things and AI, can tune in to the Earth system, and monitor its progress. Adjustments can be made in terms of our input, but it is the Earth system that will do the healing, recovery and functioning, with us as part of it. We are within the system, whether we like it or not. It's the way that it is. If we want to continue to enjoy the good fortune of dwelling on the jewel in the crown of our solar system, we need to re-integrate and embrace the four truths: non-linearity, emergence, sub-optimality and real-time feedback. Because this is what is meant by living within a system.

The path ahead will emerge whatever happens, good or bad, but if we transition in such a way as to fit back into the system, climbing on board the bus as it leaves, what emerges will include us and will take the appropriate form. We are not looking at a destination with a street plan of our making. There is no map of this land, but it is the Earth system that will produce whatever it will produce.

It is the unrolling of a scroll, true emergent evolution. And this is why this book has not sketched out the structures of transition. Rather the change must come from within the system and relate to our functioning as citizens of Earth and as partners in the Earth system.

Appropriate forms arise from functional integrity, and functional integrity emerges from integration. Society must re-integrate into the Earth system, and then we will know what we need to do. Being comes before doing. The structural diversity of life on Earth represents a dynamic ever-changing response to a much deeper set of functional principles, and it is within these principles that we find the architect and engine of diversity itself. Once we are restored to our place within the system, the appropriate forms will manifest themselves.

This is the only way. We cannot enforce our own plans on a system so complex. Furthermore, this system has all the answers we seek. We just need to listen, through the awesome technology that is AI, and the path will reveal itself. This is not some mystical, absurd ideology, but one rooted in systems theory, ecology, technology and sociology. And this is also where AI can step forward as the perfect technology at the perfect time.

While its development had been envisaged variously as a path to Enlightenment utopia, as an economic powerhouse delivering unprecedented growth and wealth or as the greatest existentialist threat to humankind, in this book we see it pull up a chair at the table in our moment of despair, and offer the hand of friendship, the invisible hand, connecting us to our planet.

AI has found a new role, more important than we could have imagined in our wildest technological fantasies. It has revived the character of guardian, a modern-day Oracle of Delphi, communicating between humans and the Earth system, and providing the guidance we need. It is the bridge across the chasm, a chasm created by humans in our obsessive, destructive search for truth within ourselves. And its portal is the internet of things, that constellation of ears and eyes across the planet, from the smart meter in your house to the satellites encircling the globe some twenty thousand miles above your head. Listening, analysing, informing, this technological Argos is the giant with seven billion eyes.

We are Midas, King of Phrygia, foolish and greedy, turning all that we touch into gold. We have wreaked havoc, transforming the Earth into material wealth. But Midas cursed his gift, when he realized that even his food and water were turned to gold, and he starved and thirsted. Midas then failed to appreciate the music of Apollo, the god of music, preferring the pipes of Pan. Apollo turned the ears of Midas into those of a donkey, proclaiming that Midas must have the ears of an ass.

And such is our story. Destroying the Earth system that nourishes us, in search of profit, and then failing to listen to the feedback until it is almost too late. But the solutions are out there, all around us, a living breathing planet capable of recovery, quite able to continue without us, but also willing to accommodate us if we change our ways fundamentally.

And the technology is there to change those big, hairy ears into instruments of attentiveness and discernment. There have been stirrings and awakenings before. Henry David Thoreau in the nineteenth century inspired many, including John Muir, the founder of the Sierra Club. *Silent Spring*, by Rachel Carson, roused environmental movements in the sixties and seventies, including the deep ecology movement.

But today we are in a time of global emergency, and there can be no turning back. We are without excuse, and have all that we need for this necessary transition. He that hath an ear, let him hear.

Glossary

Actor-network theory: A theory that emphasizes that humans and non-humans are intermeshed and that there is no single actor, but a constantly shifting network of actors. Ultimately, the interactions are everything.

Agrarian age: Period during which a group of people practiced agriculture, utilizing solar energy to power society. Different parts of the world have entered and left an agrarian phase at different times. Currently, there are many people who still live in an agrarian age. Agrarianism is preceded by a **hunter-gatherer age**, and followed by an industrial age. The onset of agrarianism is viewed by many as the onset of modernism in terms of culture, civilization and economics.

Algorithm: A series of instructions setting out how to solve a problem or reach a goal.

Anthropocene: A suggested new geological epoch, representing the time during which humans have impacted the planet in such a way as to leave a geological record, in terms of changes to weathering, sedimentation and deposition of radioactive elements. The concept has been around in different forms for 200 years, but has recently been popularized by Paul Crutzen and Eugene Stoermer.

Artificial general intelligence (AGI): Machine capable of solving any task a human could solve. AGI has the capacity for sentience, cognition and functioning potentially far in advance of what humans are capable of, referred to as superintelligence. See also **Strong AI**.

Artificial intelligence: An area of computer science that emphasizes the creation of intelligent agents, thinking, functioning and responding like humans. These agents are capable of perceiving their environment, interpreting this input, acting in such a way as to maximize successful outcomes and goal achievements and learning from the experience.

Asimov's laws: The four laws were developed as part of Isaac Asimov's science fiction writing. The first law states that a robot may not injure a human being or, through inaction, allow a human being to come to harm. The second law states that a robot must obey the orders given to it by human beings except where such orders would conflict with the first law. The third law states that a robot must protect its own existence as long as such protection does not conflict with the first or second laws. The zeroth law, added a number of years later, states that a robot may not harm humanity, or, by inaction, allow humanity to come to harm.

Associative learning: Through experience, learning in one area can be linked with learning from another area, leading to reinforcement. A simple example is Pavlov's dog, who associated a ringing bell with being fed, and, following repeated exposures to this sequence, would salivate at the sound of the bell, even though no food was on offer.

Atmosphere: The layer of gases surrounding the planet, and extending three hundred miles (four hundred and eighty kilometres) above the surface. Plays key roles in terms of protecting against ultraviolet and other ionizing radiations, providing oxygen and carbon dioxide, and keeping the planet warm, through greenhouse gases, which maintain the average temperature at 15°C rather than 0°. The atmosphere has changed through time, and provides the key medium for pollen, spore and seed distribution as well as precipitation and bulk movement of warm and cold air.

Behavioural momentum: The tendency for behaviour to persist following a change in environmental conditions. The momentum increases with increasing rates of reinforcement.

Big data: The vast array of data gathered by the internet of things. Its trend is upward, with increasing volume, variety and velocity.

Big nudging: Subtle architecting of society's choices through utilizing big data to elucidate ways of influencing society.

Biofuel: A fuel that is produced from recently fixed carbon, particularly in transportation. Ethanol is an example.

Biomimicry: A school of sustainability developed by Janine Benyus, which advocates the application of natural processes and designs to human production, aspiring to bring human activity into greater synchrony with the natural world. See **Bio-participation**.

Bio-participation: Rather than borrowing processes and designs for human use (see **Biomimicry**), bio-participation advocates the re-integration of humans within the biosphere, where participation rather than knowledge transfer ensures a deeper symbiosis.

Biosphere: The sum of all life on the planet.

Buenvivir: A school of sustainability predominately in the Andes region of South America (particularly Ecuador and Bolivia), emphasizing the importance of local geography, ecology and culture in terms of economic functioning.

Capital:

- **Human:** The human workforce in terms of numbers, skills, health and functionality.

- **Social:** The intensity, resilience, outputs and patterns of networks among people.
- **Human-made:** The material output of human activities.
- **Natural:** The functional and structural resources available in the environment.

Carbon footprint: Variously defined as the sum of all carbon dioxide or carbon-based greenhouse gases released by an individual, organization, event or nation. The footprint is the area of forest needed to absorb the said amount of carbon, but this doesn't work well for other greenhouse gases, and runs into complications in terms of what age the trees are, as this greatly effects net carbon absorption.

Carbon trading: The concept of trade of carbon emissions rights within the setting of limits on the quantity of carbon emissions allowed by a given country, person or other grouping. If your emissions are below the limited quantity, you can sell the unused quantity to someone who has reached their limit. However, the potential to reduce emissions globally is questionable, as it encourages carbon leakage, in terms of finance facilitating excess pollution. See also **Personal carbon trading.**

Carrying capacity: The number of organisms that a habitat can support without ecological degradation. This is a dynamic measure, impacted by many things.

Cerebrocentrism: The belief that only organisms with brains can be intelligent.

CFCs: Chlorofluorocarbons are gases that lead to the destruction of ozone in the atmosphere, resulting in increased UV radiation reaching the planet's surface. While emissions have greatly reduced, CFCs can function for one hundred years.

Chatbot: Artificial intelligence (AI) software that can conduct a conversation with a user in natural language (spoken or in text) using, for example, the telephone, a mobile app or a website.

Circular economy: A school of sustainability unique in terms of its incorporation into central policy by the governments of China and South Korea, with emphasis on recycling, resource use efficiency and longer life product functionality, in order to reduce the use of natural capital.

Citizen science movement: Active public involvement in scientific research. Recent advances involve smart devices.

Civil society model: Transition approach focused on empowering the community. It entails social justice, builds social capital and is rooted in traditional knowledge. It has endogenous drivers, bringing community resilience and, preferably, is based on human-environment relations, being localist in nature.

Cloud computing: Storing and accessing data and programs over the internet instead of your computer's hard drive.

Cloud robotics: The potential to have smaller, more battery-efficient robots by placing the 'brain' of the robot in the cloud. See **Cloud computing.**

Co-design: Design school that incorporates end-users in the design process, usually beyond mere feedback, and builds a synergy between designer, producer, product and customer. Emphasis is placed on the design process, rather than the designer *per se.*

Co-dissemination: To provide citizens with full access to research findings, allowing them to channel these findings into future research and development projects.

Common sense knowledge problem: The challenge of providing an AI with general knowledge that the average human would have, in order to create a thinking machine that comes across as real. This is extremely problematic, as common-sense knowledge can be deeply culture-dependent and reliant on an individual's journey of learning.

Contingency theory: Stephen Jay Gould's theory that evolution is constrained by historical events that are random in nature, so that if we filmed the story of life again it could look completely different.

Co-production: Involving citizens at the heart of a research project.

Court jester hypothesis: Emphasizes the importance of the abiotic rather than the biotic context, particularly random climate events, in determining the path of evolution. See **Red Queen Hypothesis** and **Contingency theory**.

Data-driven modelling: By analyzing the data from a particular system, models are generated based on the connectivity between input and output values, allowing models to be generated and improved upon with increasing data analysis. This very much resembles **non-symbolic programming.**

Data ethics: The branch of ethics that relates to moral problems associated with data generation, curation, analysis and use, with algorithms in artificial intelligence and with practice in terms of liability and responsibility.

Deep learning: A **machine learning** technique based on learning by example. Tasks are repeatedly carried out with small changes being made, moving toward increasingly successful outcomes, in a form of trial-and-error learning.

Deontology: A rule-book form of ethics, wherein an action is either right or wrong, and the consequences of the action are irrelevant in terms of decision-making.

Development: As applied to economics, the deliberate, funded and directed transformation of agrarian or early industrial nations into the Information Age, with emphasis on globalization and western values. Nations not yet transformed are often referred to as 'developing' while those that belong to the transformed set are referred to as 'developed' nations. See **Post-development**.

Dispersed intelligence: Separating the learning of a large system into multiple sub-systems. Particularly relevant to plants, which, without a central nervous system and brain, carry out data processing in millions of growing tips, the meristems, hence allowing different decisions to be made in different physical contexts.

Dynamic equilibrium: A school of thought that recognizes that no subsystem can reach a climax or static state, given the continuous change of the overall system (the Universe) and the ongoing interactions with other subsystems. Change is constant and results from feedback. An equilibrium represents a hypothetical state around which real sub-systems oscillate.

Earth system: The entirety of the interactions between the biosphere, atmosphere, hydrosphere and geosphere. A functioning entity made up of a number of sub-systems.

Ecological footprint: A visualization of the amount of biologically productive land and sea area necessary to replace the resources that a human population consumes, and to assimilate associated waste. Issues relate to what is considered as a reference habitat, as land and sea vary greatly in their ability to absorb waste depending on many factors.

Ecological intelligence: The ability to understand the **Earth system** and our place within it. This set of skills is thought to have weakened as we have become more distanced from the natural environment through technology.

Ecological modernization theory: Argues that continued modernization is essential for environmental sustainability. For a contra-argument, see **Treadmill of production theory.**

Ecology: The study of the **Earth system** in terms of the interactions between material flows, energetics and the biosphere. Sub-disciplines include population ecology, molecular ecology, community ecology, physiological ecology, ecosystem function and human ecology.

Economics: The study of the production, consumption and transfer of wealth.

Ecosystem: The abiotic and biotic environment of an organism, consisting of all of the interactions and representing a self-organizing unit, formed from populations of organisms and characterizing a sub-system of the Earth system.

Ecosystem intelligence: The emergent intelligence of the biosphere, arising from the interactions and feedback from all levels of organization, and leading to self-assembly and self-organization of the Earth system.

Ecosystem services: Key functional outcomes of an ecosystem that are of importance to the survival and aesthetics of organisms, including humans, within that ecosystem and in the larger biosphere.

Emergence: Outcomes which are different from the sum of the parts of a sub-system or system, such as imagination, life and ecological succession.

Emotional intelligence: The ability to monitor one's own and others' feelings and emotions, to discriminate among them and to use this information to guide one's thinking and actions.

Empiricism: A school of philosophy that asserts that knowledge is derived mostly from sensual experience rather than reasoning (rationalism). Empiricism forms the basis of the scientific method and of reductionism (where higher levels of organization are merely outcomes of lower levels).

Enlightenment: The philosophical revolution of the eighteenth century that set out a road of progress based on reason and technology, whose imagined destination would be a utopian human society. Embracing social sciences, economics, education and the natural sciences, it generally celebrated the individual and decried that which would restrain individual progress, including religion and state intervention. Nature was also viewed as an unnecessary hindrance to this progress. The Enlightenment shaped the modern world as we know it.

Environmental Kuznets curve: Curve showing the relationship between increasing environmental damage (y-axis) and increasing economic growth (x-axis). Shaped like an arch, it suggests that as the economy grows, there is an initial increase in damage, but this then levels off and decreases. Thus, economic growth is posited to deliver environmental sustainability.

Ethic of care: The answers to moral questions are seen to emerge from interpersonal interactions, rather than from the individual.

Eutrophication: Enrichment of a water body with nutrients to such a level as to destabilize ecosystem function.

Existentialism: School of philosophy emphasizing that the world is not an ordered, determined system understood through the construction of laws based on observation. Worldly pursuits are viewed as futile, and personal responsibility is key. Society is viewed as an artefact, and often a negative force. The individual should determine their own journey, unfettered by society, state or religion.

Existential risk: A threat so great as to have the potential of delivering the extinction of the human race.

Fragmented ecosystem: Refers to a lack of integration across a system, that prevents it from working as a whole. In AI, this can be because of diverse standards or technical incompatibilities.

Friluftsliv: A philosophical lifestyle based on experiences of the freedom within nature and a spiritual connectedness with the landscape.

Fuel poverty: Where family income is insufficient to keep a house adequately warm.

Fuzzy logic: Fuzzy logic deals with finding an approximate rather than a precise pattern in a data set. Rather than black or white, it explores the intermediate shades of grey, where uncertainty and incomplete data may mean dealing in degrees of truth.

Gaian hypothesis: The idea that the Earth is a self-organizing super-organism, with feedback favouring the *status quo*. A top-down control of life-processes, achieving homeostasis, sets this theory apart from reductionist theories. Named after the Greek goddess, *Gaia*. Developed by James Lovelock and Lynn Margulis.

Garden-of-Eden complex: The conservationist tendency towards recreating some perfect ecosystem from the past, usually by re-introducing species that have become locally extinct. The desire to return to some past ecology is usually disastrous, as you cannot re-construct a system through reductionist block-building approaches.

General purpose technology: Any new technological development that is important enough to have a protracted aggregate impact. Examples include electricity and the internet.

Geosphere: The crust, mantle, the outer core and the inner core of the Earth, including the soil.

Gestalt theory: The whole is other than the sum of the parts.

Globalization (economic): The ultimate free-market economy, where all nations freely trade with each other using the same economic model, under the auspices of the invisible hand, for maximum benefit. Globalization can also extend to cultural, legal and educational fields. Globalization is strongly opposed by many, who argue that diversity is essential for resilience, and that the model being used is a Western, neo-colonial one.

Greenhouse gases: Any gas that contributes to the greenhouse effect, wherein heat escaping from the planet is absorbed by greenhouse gas molecules and re-radiated back to Earth, contributing to the warming of the planet.

Gross domestic product (GDP): The monetary value of all the finished goods and services produced within a country's borders in a specific time period.

Growth (economic): A long-term expansion of a country's productive potential.

Gut-brain axis: The gut-brain axis is a bidirectional link between the central nervous system (CNS) and the gut of the body, involving direct and indirect conversations between cognitive and emotional centres in the brain with peripheral intestinal functions. The microflora in the gut play a significant role in impacting the cognitive and emotional states of the brain.

Hardware approach: In artificial intelligence, this represents efforts to mimic the structure of the central nervous system in order to attempt to arrive at human-like intelligence, wherein the components can allow the AI to think for themselves, as in the **non-symbolic AI**, rather than rely on software (**symbolic AI**) which tells the AI what to do.

High-grading: Natural resources, both renewable and non-renewable, are always harvested in such a way so that the easiest and cheapest supplies are taken first.

Horizontal gene transfer: The ability of bacteria to copy parts of their DNA (genes) and then transfer them to other bacteria, often not of the same type. This sharing of information gives bacteria a huge advantage in terms of adapting to changing environments, wherein once random mutations produce a new solution, this can quickly spread to other bacteria. Antibiotic resistance is a classic example.

Humanism: A philosophical school that emphasizes the value and agency of human beings.

Hunter-Gatherer Age: Period of time when a given group of individuals hunted and foraged for food, often in a nomadic existence, with neither settlements, economics nor agriculture. Strong feedback between the environment and human population existed, given that no food was stored. This represents the natural state of most animals on Earth. Humans have spent at least ninety-seven percent of their existence in this state, and there are still small numbers of humans who live in this way.

Hydrosphere: All of the liquid water on Earth, forming an essential resource for all life and an important habitat, where life first evolved. The water cycle is a central process of the Earth system.

Inbreeding depression: Genetic abnormalities produced by sexual reproduction between close relatives, leading to phenotypic abnormalities that often prevent reproduction, such as egg casings in birds which are so thin, the egg cannot survive.

Infosphere: The world of communication, where information is gathered, exchanged, analyzed and used. A particular emphasis on digital communication, wherein the digital age has created a parallel universe, in which we are represented as avatars of ourselves. The infosphere and real world are becoming intermeshed.

Institutional silo: An attitude where several departments or groups do not want to share information or knowledge with other individuals in the same organization, leading to a reduced likelihood of resolving complex problems.

Instrumentalist technology: The viewpoint that technology is neutral and can be transferred from industrialized to 'developing' worlds without any difficulty.

Intermediate disturbance hypothesis: The concept that diversity is maximum at levels of disturbance that are not too high nor low. It has been applied to many aspects of evolutionary biology (mutation rates), ecology (species richness) and thermodynamics (maximum entropy production principle).

Internet of things: The interconnected billions of **smart devices**, all gathering and distributing vast amounts of data through the internet. These smart devices are becoming increasingly invisible, and are found throughout our world, providing the eyes and ears of the **infosphere**.

Intersectionality: A framework of interacting experiences, identities and influences that combine to provide the context of prejudice and disadvantage of a person, society or ecosystem. Also used in terms of the multidimensionality, complexity and interactivity that should be considered when approaching interpretation, problem-solving and decision-making.

Invisible hand: Much debated concept, developed by Adam Smith, wherein free-market economies would be controlled by an invisible hand, which is thought to reference moral sentiments within individuals.

Kuznets curve: A graph with increasing social inequality along the y-axis and increasing economic growth along the x-axis. With increasing economic growth, inequality first increases to a maximum, but then decreases. This forms the shape of an arch. The message is that economic growth will, ultimately, deliver social equality.

Laissez-faire economics: As set out by Adam Smith, a globalized economy should be allowed to operate without interference (*laissez-faire* is a French term meaning to leave alone). See **Invisible hand**.

Liebig's law of the minimum: If one of the essential plant nutrients is deficient, plant growth will be poor even when all other essential nutrients are abundant. Also, if the limiting nutrient is increased in supply, the previously next most limiting nutrient will then become the limiting factor. More broadly applied to problem-solving, where all contributing factors must be examined.

Localism: A school of sustainability that places community and sustainability over appropriate technology. An anarchic movement, opposing centralization, it has flourished in many different forms throughout human history.

Machine learning: Enabling digital technology to act and learn like humans do, improving their learning over time in autonomous fashion. This is achieved by providing a flow of data and information in the form of observations and real-world interactions.

Macro-feedback: Part of the three key interactions between AI and humans suggested in this book in terms of delivering societal transition, the others being **micro-feedback** and **meso-feedback**. Macro-feedback represents data on the planet's health, gathered from remote sensing from satellites and land-based devices, and can be used to follow changes in the Earth system, allowing us to be aware of the impacts of our activities, good and bad, upon the planet.

Marxism: The system of socialism of which the dominant feature is public ownership of the means of production, distribution, and exchange. Class struggle is a central element, with revolution being called for to topple the Capitalist system, leading to a classless, self-governing society.

Mass extinction: The extinction of a large number of species within a relatively short period of geological time, thought to be due to factors such as a catastrophic global event or widespread environmental change that occurs too rapidly for most species to adapt. There have been five such events in the history of life on Earth, and some suggest that we are in the midst of a sixth event, caused by humans.

Meso-feedback: Part of the three key interactions between AI and humans suggested in this book in terms of delivering societal transition, the others being **micro-feedback** and **macro-feedback.** Meso-feedback targets communities, in terms of their sustainability efforts and in terms of catalysing community empowerment, cohesion and interconnectivity. Functional, integrated communities are viewed as central to any attempt at re-integrating into the Earth system and therefore sustaining our future participation.

Micro-feedback: Part of the three key interactions between AI and humans suggested in this book in terms of delivering societal transition, the others being **meso-feedback** and **macro-feedback**. Feedback to individuals relating to their resource use, damage to the environment and positive activities, based upon energy use, supply chain consequences and ecological footprints. A key phase in terms of accountability and in terms of overcoming psychological barriers to change.

Millennium Development Goals: Developed by the United Nations and running from 2000 to 2015, the MDGs focused on poverty as the central issue, using private sector funding to target a trickle-down approach, wherein raising economic standards would lead to concomitant improvements in health and education. These goals were widely criticised and viewed as irrelevant and condescending.

Morgan's canon: Higher-order explanations should never be invoked to explain behaviour when a simpler one will suffice.

Multiple Intelligence theory: This theory sets out eight different types of intelligence: linguistic, logical-mathematical, spatial, bodily/kinesthetic, musical, interpersonal, intrapersonal, and naturalistic. An individual who has a particular strength in one type of intelligence will not necessarily demonstrate a comparable aptitude in another intelligence type and therefore IQ testing is not a fair comparison of intellectual ability. The theory also posits that formal education programmes do not adequately address the spectrum of intelligence types within a given population, favouring individuals with particular strengths.

Neo-Darwinism: Darwinian evolution embracing the modern evolutionary synthesis and focused on the gene as the unit of selection and as selfish. A strongly reductionist theory.

Nihilism: A philosophy that holds that nothing in the world has a real existence and that life lacks intrinsic purpose or meaning.

Non-linearity: Describes a situation where the changes of the output are not proportional to the changes of the input. One thing does not necessarily follow another and cause-and-effect breaks down because of unknowns in the system.

Non-symbolic AI: Rather than providing fixed rules as to how to handle data, this approach involves providing raw environmental data to the machine and then the machine is left to recognize patterns, creating its own complex representations of the raw data being provided to it. Reliance is on the hardware (a neural network) rather than the software (a programme). Machine learning allows the AI to develop its own strategies and the exact process is unverifiable and uncharacterizable.

Nudging: Subtle, positive re-enforcement designed to alter the behaviour of an individual by convincing them that they, themselves, have made the decision without external interference. Usually uses previous data on the person in order to select weaknesses or biases in their mentality based on, for example, previous choices, in order to nudge them in a particular direction. Nudges have been used in advertising campaigns and elections. Nudging at a societal level is called **Big nudging.**

Ogiek people: A non-nomadic hunter gatherer tribe mostly living in the Mau forest in the Great Rift Valley in Kenya, whose economy is based around honey.

Panspermia: The theory that life originated beyond Earth, and spread here via spores transported on asteroids or other material from space.

Personal carbon trading (PCT): A concept that involves individuals being given a certain quantity of carbon pollution that they are allowed to produce, and if they do not use all of this, they can trade the spare capacity.

Plasmid transfer: The transfer of genetic material between bacteria. See also **Horizontal gene transfer.**

Pluriverse: The concept that different ecological contexts will demand different societal functioning and different economic models. Contrasts with **globalization.**

Post-development: A school of sustainability that does not call for an alternative form of development, but rather an eradication of any attempt to train the '*developing*' world, arguing that indigenous people are likely to have better solutions within the context of their natural and cultural landscape for the problems facing them than will Western thinkers.

Potemkin village: An impressive facade or show designed to hide an undesirable fact or condition.

Primary succession: The process where an ecosystem develops from a bare substrate such as a sandy shore, passing through a number of stages before developing into a dynamic but stable point, such as a forest. Succession is directional and predictable. Species are replaced by other species throughout the process.

Progress: The Enlightenment mantra that humans continue to improve and that there is nothing to be learned from our imperfect past. Rather, reasoning and technology will continue to deliver an ever-improving context for human existence.

Quorum sensing: The regulation of gene expression in response to fluctuations in cell-population density. Bacteria release chemicals which, when the population is of a certain size, exceed a threshold, often leading to a behavioural change.

Red Queen hypothesis: Organisms must continuously adapt to survive, because other organisms are changing. This amounts to an evolutionary dance through time, with evolution being driven by biotic interactions. Opposing views include **Contingency theory** and the **Court jester hypothesis.**

Relativism: Truth, knowledge and morality are not absolute nor universal, but rather are dependent on context.

Resilience: The ability to bounce back and recover from disturbance.

Resistance: The ability to continue in spite of challenges. In ecology, resistance represents a population remaining unchanged in spite of disturbance.

Scientific determinism: Relying on cause-and-effect, this philosophy dictates that everything is predetermined and an outcome of causes, and thus human acts and choices are also determined. There is no such thing as free will, as everything is controlled by internal and external forces beyond our control.

Secondary succession. The development of an ecosystem following a significant disturbance, such as a fire or volcanic explosion, where the entire ecosystem collapses and must rebuild. This differs from **primary succession** as it is a reboot, and usually there is some remnant of the previous ecosystem, such as soil, seeds or rootstocks. Thus, it is not an identical journey compared to primary succession.

Silo: See **institutional silo**.

Situated intelligence: The theory that sets out the idea of intelligence emerging from the close interaction between the individual and their environment, wherein any understanding of our intelligence must rest on the fact that we are situated within that environment.

Situation ethics: The theory that proposes that ethical decision-making should be flexible and depend on the circumstances.

Smart device: An electronic device that is connected to other devices or networks and forms part of the **internet of things**.

Social intelligence: The ability to navigate social landscapes and to build appropriate relationships.

Socio-metabolic profile: Social structures and their use of resources (material and energy).

Software approach: An approach to AI where the processes are programmed into the machine and so outputs are predictable from inputs. Thus, cause-and-effect dominates and there is a predictable, verifiable and controlled process. This is the basis of **symbolic AI** and is very different than the **hardware approach** in **non-symbolic AI**.

Soil salinization: The build-up of salt in the soil, accumulated from irrigation, where tiny amounts of salt in fresh water gradually build up, decimating agricultural productivity. This process is extremely difficult to reverse. Salt may also rise to the surface from soils formed over ancient seas. Here the salt is brought to the surface from saline groundwater aquifers receiving too much rainwater and rising, often because trees have been cut down, reducing the amount of water being taken up by the trees.

Species redundancy: Where a number of species play the same functional role in an ecosystem, meaning that if one of the species went extinct, other species could still maintain ecosystem function. This redundancy provides resilience against species loss.

Stigmergy: A mechanism of indirect coordination, wherein the trace left by an action in a medium stimulates subsequent actions. For example, pheromone trails left by one individual can impact the behaviour of others. This can lead to complex, co-ordinated group behaviour emerging without the need for planning or control.

Strong AI: Meaning the same as **artificial general intelligence** (AGI), strong AI is a form of AI that can think like a human and be equal in all intellectual capacity and performance. Capabilities should include problem-solving, communication, planning, strategizing, decision-making, improvement and sentience. Of course, since humans vary in their intellectual capacities and ways of thinking, it is unclear exactly which human would be compared to the machine as a standard.

Strong sustainability: A school of sustainability that states that natural capital must be protected at all costs and cannot be replaced with human made capital. It sees little role for technology in any sustainability discourse.

Sub-optimality: Since each level of organization is part of another level, and so solution space will be compromised, leading to reduced optimality for any given level, but overall optimality at the system level.

Substantive technology: Advocates that technological development exists within its own bubble, one invention building on the other, without anyone or any institution being able to stop, slow or redirect it. From this perspective, technology is not neutral.

Succession: The process of ecosystem development, from bare soil or sand to forest in the example of lowland succession. See **Primary succession** and **Secondary succession**.

Supply chain: All of the material, environmental and social players involved in the production of a product, and the impacts of such processes on society, economics and the environment. Supply chains today are often global in their geography, given the complexity of many manufactured goods. **Ecological footprints** are associated with such chains.

Sustainability: The maintenance of capital. Social sustainability maintains human capital, economic sustainability maintains human-made capital and environmental sustainability maintains natural capital. Two more general forms exist: **strong sustainability** and **weak sustainability**.

Sustainable development: Development that meets the needs of the present without compromising the ability of future generations to meet their own needs.

Sustainable development goals: A United Nations development programme, running from 2015 to 2030, which has seventeen goals, one hundred and sixty-nine targets and two hundred and thirty-two indicators, addressing social, economic and environmental sustainability issues. Incorporating a systems approach and designed by a wide range of contributors from government to civil society, they are looked on as far superior to their predecessor, the **Millennium development goals**.

Swarm intelligence: The situation where individuals gain information independently but then combine this information to create a solution, overcoming each individual's cognitive limitations.

Symbolic AI: Real world entities are represented with symbols and problems are solved by using these symbols in a logical way. The AI is programmed with exact instructions and rules as to how to manage these symbols in an if/then approach. The outputs are completely predictable from the inputs, as the processes in between are determined by the human programmer.

Systems theory: The transdisciplinary study of complex, multi-component systems, emphasizing non-linearity, emergence, sub-optimality and feedback. The emphasis is on holistic, not reductionist approaches, and works within incomplete knowledge. Non-symbolic AI is understood within systems theory, wherein outputs emerge from a black box of interactions.

Top-down model: A process of transition where institutions imprint solutions upon society. Here, governments take control of the process, and enforce change through regulations and policies.

Trade off: When two or more challenges exist relating to a problem, it is impossible to optimize the solution for all challenges. Trade-offs are required in order to achieve an overall solution. For example, a station wagon will not be as aerodynamic as a sports car, but will have more space for luggage. Compromise is important if a resolution is to be found.

Tragedy of the commons: The observation that if a resource of limited quantity is accessed by a population, then there is a likelihood that the resource will be over-exploited, to the detriment of the population, because individuals will continue to use the resource with no concern for the greater population.

Transformation: A radical change or alteration.

Transition: A transformation process in which existing structures, institutions, culture and practices are broken down and new ones are established.

Treadmill of production theory: This theory emphasizes that advanced economies are trapped on a treadmill, wherein the need for continued economic growth does not improve human well-being but has an increasingly negative impact on the environment, which in turn negatively impacts human well-being.

Uncertainty principle: This principle states that the speed and position of the tiny particles that make up our atoms cannot both be known at the same time. Also, any attempt made to measure these two properties will alter the object being observed (the observer effect).

Utilitarianism: A school of ethics wherein outcomes govern whether or not something is right or wrong. This represents the opposite position to that held by **deontology.**

Virtue ethics: A school of ethics based on individuals and their intentions. Individuals require moral education to help them determine what is virtuous and what is not. An action is viewed as being morally right if the actor is of morally good character.

Water footprint: An empirical indicator of how much water is consumed, when and where, measured over the whole supply chain of the product. The water footprint of an individual, community or business is defined as the total volume of freshwater that is used to produce the goods and services consumed by the individual or community or produced by the business.

Weak AI: Machine intelligence limited to a narrow field. For example, an AI might be able to play chess but be unable to do anything else. However, it can play chess very well, unhindered by having to do other things.

Weak sustainability: School of sustainability that seeks to maintain levels of total capital from generation to generation, through substitution. Thus, if natural capital declines, provided that man-made capital increases by the same amount, then total capital will be maintained.

Wicked problems: Cultural or social problems that resist resolution, due to barriers, complexity, lack of knowledge and the failure to recognize the need for trade-offs. Wicked problems are often wicked because of the failure of us to recognize that we need systems thinking to understand them.

References

Abbasi, J. (2018). Shantanu Nundy, MD: The human diagnosis project. Jama 319(4): 329-331.

Abit, M.J.M., Arnall, D.B. and Phillips, S.B. (2018). Environmental implications of precision agriculture. *In*: Shannon, D.K., Clay, D.E. and Kitchen, N.R. (eds.), Precision Agriculture Basics. Madison, WI: ASA, CSSA, and SSSA, pp. 209-220.

Accenture, P.L.C. (2017). How companies are reimagining business processes with IT. Available at https://sloanreview.mit.edu/article/will-ai-create-as-many-jobs-as-it-eliminates/ (last accessed June 24, 2019).

Acemoglu, D. and Restrepo, P. (2018). The race between man and machine: Implications of technology for growth, factor shares, and employment. American Economic Review 108(6): 1488-1542.

Agar, N. (2016). Don't worry about superintelligence. Journal of Evolution and Technology 26(1): 73-82.

Aghion, P., Jones, B.F. and Jones, C.I. (2017). Artificial Intelligence and Economic Growth (No. w23928). Stamford Institute for Economic Policy Research, Stanford University. Available at https://siepr.stanford.edu/sites/default/files/publications/17-027.pdf (last accessed June 24, 2019).

Agre, P. and Chapman, D. (1987). Pengi: An implementation of a theory of activity. Seattle, WA: AAAI-87, pp. 268-272.

Altschuler, E.L., Calude, A.S., Meade, A. and Pagel, M. (2013). Linguistic evidence supports date for Homeric epics. BioEssays 35(5): 417-420.

Aman, M.M., Solangi, K.H., Hossain, M.S., Badarudin, A., Jasmon, G.B., Mokhlis, H., Bakar, A.H.A. and Kazi, S.N. (2015). A review of safety, health and environmental (SHE) issues of solar energy systems. Renewable and Sustainable Energy Reviews 41: 1190-1204.

Ambrosetti, N. (2011). Cultural Roots of Technology: An Interdisciplinary Study of Automated Systems from the Antiquity to the Renaissance. Doctoral thesis, Università Degli Studi di Milano, Di Scienze Matematiche, Fisiche E Naturali, Dipartimento Di Informatica E Comunicazione.

Angwin, J., Larson, J., Mattu, S. and Kirchner, L. (2016). Machine bias: There's software used across the country to predict future criminals and it's biased against blacks. ProPublica, May 23, 2016. Available at https://www.propublica.org/article/machine-bias-risk-assessments-in-criminal-sentencing (last accessed June 24, 2019).

Armey, D. and Kibbe, M. (2010). A Tea Party manifesto. Wall Street Journal 17th August, 2010. Available at https://go.shoreline.edu/gac/documents/coffeecurrents/teapartymanifesto081710.pdf (last accessed June 24, 2019).

Arntz, M., Gregory, T. and Zierahn, U. (2017). Revisiting the risk of automation. Economics Letters 159: 157-160.

Babidge, S. (2016). Contested value and an ethics of resources: Water, mining and indigenous people in the Atacama Desert, Chile. The Australian Journal of Anthropology 27(1): 84-103.

Baccolini, R. and Moylan, T. (2003). Dark Horizons. Science Fiction and the Dystopian Imagination. London: Routledge.

Bacon, F. (2010). Aphorisms Concerning the Interpretation of Nature. The New Organon: Or True Directions Concerning the Interpretation of Nature. Book 1: 6. Available at http://www.earlymoderntexts.com/assets/pdfs/bacon1620.pdf (last accessed June 24, 2019).

Baker, C. (2018). Programming today's AI for tomorrow's gender equality. Computer Business Review 18th May, 2018. Available at https://www.cbronline.com/opinion/programming-todays-ai-tomorrows-gender-equality/ (last accessed June 24, 2019).

Banaszak, A.T. and Lesser, M.P. (2009). Effects of solar ultraviolet radiation on coral reef organisms. Photochemical and Photobiological Sciences 8(9): 1276-1294.

Bano, M. (2018). Artificial intelligence is demonstrating gender bias – and it's our fault. Kings College, London New Centre, 25th July, 2018. Available at https://www.kcl.ac.uk/news/artificial-intelligence-is-demonstrating-gender-bias-and-its-our-fault (last accessed June 24, 2019).

Banuri, T., Weyant, J., Akumu, G., Najam, A., Rosa, L.P., Rayner, S., Sachs, W., Sharma, R. and Yohe, G. (2001). Setting the stage: Climate change and sustainable development. *In*: Metz, B., Davidson, O., Swart, R. and Pan, J. (eds.), Climate Change 2001: Contribution of Working Party III to the Third Assessment Report of the Intergovernmental Panel on Climate Change. Cambridge: Cambridge University Press, pp. 73-114.

Barkin, D. (2010). Incorporating indigenous epistemologies into the construction of alternative strategies to globalization to promote sustainable regional resource management: The struggle for local autonomy in a multi-ethnic society. *In*: Esquith, S. and Gifford, F. (eds.), Capabilities, Power and Institutions. Towards a More Critical Development Ethics. University Park, PA: Penn State University Press, pp. 142-161.

Barnes, E. (2016). Advanced Artificial Intelligence: Policy and Strategy. CUSPE Communications. Available at https://www.repository.cam.ac.uk/bitstream/handle/1810/278275/Barnes-Oct-2016-2.pdf?sequence=1 (last accessed June 24, 2019).

Barnes, S. (2018). There are two types of companies: Those who know they've been hacked and those who don't. Dynamic Business. 29 March 2018. Available at https://www.dynamicbusiness.com.au/technology/there-are-two-types-of-companies-those-who-know-theyve-been-hacked-those-who-dont.html (last accessed June 24, 2019).

Barnett, L. (1949). The Universe and Dr Einstein. London: The Fanfare Press.

Barnosky, A.D. (1999). Does evolution dance to the red queen or the court jester? Journal of Vertebrate Paleontology 19: 31A.

Barrat, J. (2013). Our Final Invention: Artificial Intelligence and the End of the Human Era. New York: Thomas Dunne Books.

Bartels, D. and Sunkar, R. (2005). Drought and salt tolerance in plants. Critical Reviews in Plant Science 24: 23-58.

Bastiat, F. (1964). Selected Essays on Political Economy. Irvington-on-Hudson, NY: Foundation for Economic Education, pp. 2-4.

Berger, P.L. and Neuhaus, R.J. (1977). To Empower People: The Role of Mediating Structures in Public Policy. Washington DC: American Enterprise Institute for Public Policy Research.

Bernardi, G. (2012). The use of tools by wrasses (Labridae). Coral Reefs 31(1): 39.

Berry, W. (2012, originally, 1967). The Long-Legged House. Berkeley: Counterpoint.

Blackburn, R. (1970). A preliminary report of research on the Ogiek Tribe of Kenya. Discussion paper No. 89. Institute for Development Studies, University College Nairobi. Available at https://opendocs.ids.ac.uk/opendocs/bitstream/handle/123456789/441/dp89-317999.pdf?sequence (last accessed June 24, 2019).

Bogost, I. (2015). The cathedral of computation. Atlantic, January 15, 2015. Available at http://www.theatlantic.com/technology/archive/2015/01/the-cathedral-of-computation/384300/ (last accessed June 24, 2019).

Bonnefon, J.-F., Shariff, A. and Rahwan, I. (2016). The social dilemma of autonomous vehicles. Science 352(6293): 1573-1576. Available at https://arxiv.org/ftp/arxiv/papers/1510/1510.03346.pdf (last accessed June 24, 2019).

Boring, E.G. (1923). Intelligence as the tests test it. New Republic 35: 35-37.

Bosman, L.A. (2003). Correlates and Outcomes of Emotional Intelligence in Organizations. Faculty of Economic and Building Sciences. Doctoral Thesis. Port Elizabeth: University of Port Elizabeth.

Bostrom, N. (2003). Ethical issues in advanced artificial intelligence. *In*: Lasker, G.E., Wallach, W. and Smit, I. (eds.), Cognitive, Emotive and Ethical Aspects of Decision Making in Humans and in Artificial Intelligence, Vol. 2. International Institute of Advanced Studies in Systems Research and Cybernetics, pp. 12-17.

Bostrom, N. and Yudkowsky, E. (2014). The ethics of artificial intelligence. The Cambridge Handbook of Artificial Intelligence 1: 316-334.

Bourguignon, F. (2017). The Globalization of Inequality. Princeton, New Jersey: Princeton University Press.

Bouton, C.E., Shaikhouni, A., Annetta, N.V., Bockbrader, M.A., Friedenberg, D.A., Nielson, D.M., Sharma, G., Sederberg, P.B., Glenn, B.C., Mysiw, W.J. and Morgan, A.G. (2016). Restoring cortical control of functional movement in a human with quadriplegia. Nature 533(7602): 247-250.

Bowler, C. and Chua, N.H. (1994). Emerging themes of plant signal transduction. The Plant Cell 6(11): 1529-1541.

Brewster, T. (2018). A hacker forced 50,000 printers to spread PewDiePie propaganda – and the problem is much bigger than you know. Forbes, 3rd December, 2018. Available at https://www.forbes.com/sites/thomasbrewster/2018/12/03/a-hacker-forced-50000-printers-to-spread-pewdiepie-propagandaand-the-problem-is-much-bigger-than-you-know/#675217293819 (last accessed June 24, 2019).

British Scence Association (2015). One in three believes that the rise of artificial intelligence is a threat to humanity. Retrieved September 21, 2016. Available at http://www.britishscienceassociation.org/news/rise-of-artificial-intelligence-is-a-threat-to-humanity (last accessed June 24, 2019).

Brooks, R.A. (1990). Elephants don't play chess. Robotics and Autonomous Systems 6(1-2): 3-15.

Brooks, R.A. (2014a). How to build complete creatures rather than isolated cognitive simulators. *In*: VanLehn, K. (ed.), Architectures for Intelligence. Hillsdale, New Jersey: Lawrence Erlbaum Associates, Inc., pp. 225-241.

Brooks, R.A. (2014b). Artificial intelligence is a tool, not a threat. Rethink Robotics (blog).

Brown, A. (2010). The saddest farewell: How the harsh way of life on St Kilda came to an end 80 years ago. Daily Record, 21st August, 2010. Available at https://www.dailyrecord.co.uk/news/uk-world-news/the-saddest-farewell-how-the-harsh-way-1067790 (last accessed June 24, 2019).

Brown, L.R. (2011). The new geopolitics of food. Food and Democracy 23. Available at http://www.foreignpolicy.com/articles/2011/04/25/the_new_geopolitics_of_food (last accessed June 24, 2019).

Brown, D. (2018) Google Diversity Annual Report 2018. Google Inc.

Brown, D. and Parker, M. (2019). Google Diversity Annual Report. Available at https://diversity.google/annual-report/ (last accessed June 24, 2019).

Brown, N. and Sandholm, T. (2018). Superhuman AI for heads-up no-limit poker: Libratus beats top professionals. Science 359(6374): 418-424.

Brown, C., Laland, K. and Krause, J. (2011). Fish cognition and behavior. *In*: Brown, C., Krause, J. and Laland, K. (eds.), Fish Cognition and Behaviour. Oxford: Wiley, pp. 1-9.

Burke, E. (1970, first published in 1759). A Philosophical Enquiry into the Origin of our Ideas of the Sublime and the Beautiful. Second Edition. Meston: Scolar Press.

Burke, M., Davis, W.M. and Diffenbaugh, N.S. (2018). Large potential reduction in economic damages under UN mitigation targets. Nature 557: 549-553.

Brynjolfsson, E. and McAfee, A. (2011). Race against the machine: How the digital revolution is accelerating innovation, driving productivity, and irreversibly transforming employment and the economy. Digital Frontier Press. Available at http://digital.mit.edu/research/briefs/brynjolfsson_McAfee_Race_Against_the_Machine.pdf (last accessed June 24, 2019).

Callon, M. (1999). Actor-network theory – the market test. The Sociological Review 47(1_suppl): 181-195.

Calvo Garzón, P. and Keijzer, F. (2011). Plants: Adaptive behavior, root-brains, and minimal cognition. Adaptive Behavior 19(3): 155-171.

Cantrell, B., Martin, L.J. and Ellis, E.C. (2017). Designing autonomy: Opportunities for new wildness in the Anthropocene. Trends in Ecology and Evolution 32(3): 156-166.

Carabotti, M., Scirocco, A., Maselli, M.A. and Carola, S. (2015). The gut-brain axis: Interactions between enteric microbiota, central and enteric nervous systems. Annals of Gastroenterology 28: 203-209.

Carney, M. (2018). New economy, new finance, new bank. Mansion House Speech. 21st June, 2018, p. 5. Available at https://www.bankofengland.co.uk/-/media/boe/files/speech/2018/new-economy-new-finance-new-bank-speech-by-mark-carney (last accessed June 24, 2019).

Case, A. and Deaton, A. (2017). Mortality and morbidity in the 21st century. Brookings Papers on Economic Activity. Spring 2017: 397-476.

CEC (2000). Communication from the Commission on the precautionary principle. COM (2000) final. Brussels, Belgium: Commission of the European Communities.

Cellan-Jones, R. (2014). Stephen Hawking warns artificial intelligence could end mankind. BBC News, December 2. Available at www.bbc.com/news/technology-30290540 (last accessed June 24, 2019).

CERNA Opinion (2014). Éthique de la recherche enrobotique. (In French). Allistene, Paris, France. Available at http://cerna-ethics-allistene.org/digitalAssets/38/38704_Avis_robotique_livret.pdf (last accessed June 24, 2019).

Ceuppens, J. and Wopereis, M.C.S. (1999). Impact of non-drained irrigated rice cropping on soil salinization in the Senegal River Delta. Geoderma 92(1-2): 125-140.

Chakraborty, A. and Kar, A.K. (2017). Swarm intelligence: A review of algorithms. *In*: Patnaik, S. (ed.), Nature-Inspired Computing and Optimization. Cham, Switzerland: Springer, pp. 475-494.

Chella, A., Coradeschi, S., Frixione, M. and Saffiotti, A. (2004). Perceptual anchoring via conceptual spaces. *In*: Coradeschi, S. and Saffiotti, A. (eds.), Proceedings of the AAAI-04 Workshop on Anchoring Symbols to Sensor Data. Menlo Park, CA: The AAAI Press, pp. 40-45.

Chivers, D.P., Brown, G.E. and Smith, R.J.F. (1995). Familiarity and shoal cohesion in fathead minnows (*Pimephales promelas*) – implications for antipredator behavior. Canadian Journal of Zoology 73: 955-960.

Citron, D.K. (2007). Technological due process. Washington University Law Review 85(6): 1249-1313.

Clare, C. (2008). City of Ashes. London: Walker Books.

Clash, J. (2016). Formula One legend Sir Jackie Stewart on dyslexia, being bullied and more. Forbes Media LLC. 25th August, 2016. Available at https://www.forbes.com/sites/jimclash/2016/08/25/dyslexic-formula-one-legend-sir-jackie-stewart-on-being-bullied-more/#58d159e55b23 (last accessed June 24, 2019).

Clayton, S., Manning, C.M., Krygsman, K. and Speiser, M. (2017). Mental Health and Our Changing Climate: Impacts, Implications, and Guidance. Washington, D.C.: American Psychological Association and ecoAmerica.

Clifford, C. (2017). Hundreds of A.I. experts echo Elon Musk, Stephen Hawking in call for a ban on killer robots. CNBC, November 8. Available at https://www.cnbc.com/2017/11/08/ai-experts-join-elon-musk-stephen-hawking-call-for-killer-robot-ban.html (last accessed June 24, 2019).

Clogg, R. (1979). An attempt to revive Turkish printing in Istanbul in 1779. International Journal of Middle East Studies 10(1): 67-70.

Condorcet, M.J. A de (1955, originally 1779). Sketch for a Historical Picture of the Progress of the Human Mind. Translated by June Barraclough. London: Weidenfeld and Nicolson.

Connell, R. (1887). St Kilda and the St Kildians. London: Hamilton and Adams.

Conway, G.R. and Pretty, J.N. (1991). Unwelcome Harvest: Agriculture and Pollution. Earthscan Publications, London.

Conway, J. and Singh, J. (2011). Radical democracy in global perspective: Notes from the pluriverse. Third World Quarterly 32(4): 689-706.

Corbett, C. and Hill, C. (2015). Solving the equation: The variables for women's success in engineering and computing. The American Association of University Women, March 2015. Available at http://www.aauw.org/files/2015/03/Solving-the-Equation-report-nsa.pdf (last accessed June 24, 2019).

Cordell, D., Drangert, J.O. and White, S. (2009). The story of phosphorus: Global food security and food for thought. Global Environmental Change 19(2): 292-305.

Cowen, T. (2018). Neglected open questions in the economics of artificial intelligence. *In*: Agrawal, A.K., Gans, J. and Goldfarb, A. (eds.), The Economics of Artificial Intelligence: An Agenda. Chicago: University of Chicago Press, pp. 391-395.

Cox, S.J.B. (1985). No tragedy of the commons. Environmental Ethics 7(1): 49-61.

Crawford, K. (2016). Artificial intelligence's white guy problem. The New York Times, 26th June, 2016. Available at https://www.nytimes.com/2016/06/26/opinion/

sunday/artificial-intelligences-white-guy-problem.html (last accessed June 24, 2019).
Cross, F.B. (1996). Paradoxical perils of the precautionary principle. Washington and Lee Law Review 53: 851-925. Available at https://scholarlycommons.law.wlu.edu/cgi/viewcontent.cgi?article=1656&context=wlulr (last accessed June 24, 2019).
Cunfer, G. (2008). Scaling the dust bowl. Placing history: How maps, spatial data and GIS are changing historical scholarship. Redlands, CA: ESRI Press, pp. 95-121. Available at https://esripress.esri.com/storage/esripress/images/133/knowles.pdf (last accessed June 24, 2019).
Dabla-Norris, E. and Kochhar, K. (2018). Women, technology, and the future of work. IMF blog, Insights and Analysis on Economics and Finance. 16th November, 2018. Available at https://blogs.imf.org/2018/11/16/women-technology-and-the-future-of-work/ (last accessed June 24, 2019).
Danaher, J. (2015). Why AI doomsayers are like sceptical theists and why it matters. Minds and Machines 25(3): 231-246.
Darrach, B. (1970). Meet Shaky, the first electronic person. Life Magazine 20th November 1970. Available at https://books.google.co.uk/books?id=2FMEAAAAMBAJ&pg=PA71&source=gbs_toc_r&redir_esc=y#v=onepage&q&f=false (last accessed June 24, 2019).
Darwin, C.R. (1859). On the Origin of Species by Means of Natural Selection, or the Preservation of Favoured Races in the Struggle for Life. London: John Murray.
Darwin, C.R. (1994, originally 1859). On the Origin of Species by Means of Natural Selection, or the Preservation of Favoured Races in the Struggle for Life. London: Senate.
Darwin, C.R. (1880). The Power of Movements in Plants. London: John Murray.
Davies, J.S. (2008). Double-devolution or double-dealing? The local government white paper and the Lyons review. Local Government Studies 34(1): 3-22.
Dawkins, C.R. (1982). The Extended Phenotype. Oxford: Oxford University Press.
De Cremer, D., Snyder, M. and Dewitte, S. (2001). 'The less I trust, the less I contribute (or not)?' The effects of trust, accountability and self-monitoring in social dilemmas. European Journal of Social Psychology 31(1): 93-107.
DeDeo, S. (2015). Wrong Side of the Tracks: Big Data and Protected Categories. Ithaca, N.Y.: Cornell University Library, May 28, 2015. Available at https://arxiv.org/pdf/1412.4643v2.pdf (last accessed June 24, 2019).
Dehaan, R.L. and Taylor, G.R. (2002). Field-derived spectra of salinized soils and vegetation as indicators of irrigation-induced soil salinization. Remote Sensing of Environment 80(3): 406-417.
De Haan, J. and Rotmans, J. (2007). Pillars of change: A theoretical framework for transition models. In symposium Modelling Transitions to Sustainability at the ESEE 2007 Conference Integrating Natural and Social Sciences for Sustainability.
Delaney, D.M. (2005). What to do in a failing civilization. Proceeding of CACOR 3(6): 16-21. Available at http://www.zo.utexas.edu/courses/THOC/Readings/Delaney_Civilization.pdf (last accessed June 24, 2019).
Demertzis, K., Anezakis, V.D., Iliadis, L. and Spartalis, S. (2018). Temporal modelling of invasive species' migration in Greece from neighboring countries using fuzzy cognitive maps. *In*: MacIntyre, J., Maglogiannis, I., Iliadis, L. and Pimenidis, E. (eds.), IFIP International Conference on Artificial Intelligence Applications and Innovations. Cham, Switzerland: Springer, pp. 592-605.

Dempsey, N., Bramley, G., Power, S. and Brown, C. (2011). The social dimension of sustainable development: Defining urban social sustainability. Sustainable Development 19: 289-300.

Denny, M. (2017). 18 things more dangerous than sharks. PADI. Available at https://www2.padi.com/blog/2017/05/04/18-things-dangerous-sharks/ (last accessed June 24, 2019).

Dent, M. and Peters, B. (2019). The Crisis of Poverty and Debt in the Third World. Abingdon, Oxford: Routledge.

Dewey, J. (1909). Democracy and Education: An Introduction to the Philosophy of Education. New York: WLC Books.

Diekmann, A. and Preisendörfer, P. (1998). Environmental behavior: Discrepancies between aspirations and reality. Rationality and Society 10(1): 79-102.

Dietz, T., Rosa, E.A. and York, R. (2012). Environmentally efficient well-being: Is there a Kuznets curve? Applied Geography 32(1): 21-28.

Dinan, T.G., Stanton, C. and Cryan, J.F. (2013). Psychobiotics: A novel class of psychotropic. Biological Psychiatry 74: 720-726.

Dodds, W.K., Bouska, W.W., Eitzmann, J.L., Pilger, T.J., Pitts, K.L., Riley, A.J., Schloesser, J.T. and Thornbrugh, D.J. (2009). Eutrophication of US freshwaters: Analysis of potential economic damages. Environmental Science and Technology 43(1): 12-19.

Dominguez-Faus, R., Powers, S.E., Burken, J.G. and Alvarez, P.J. (2009). The water footprint of biofuels: A drink or drive issue? Environmental Science and Technology 43(9): 3005-3010.

Donne, J. (1839). The Works of John Donne. Volume III. Alford, H. (ed.). London: John W. Parker.

Draper, H. and Sorell, T. (2017). Ethical values and social care robots for older people: An international qualitative study. Ethics and Information Technology 19(1): 49-68.

Dreyfus, H.L. (1965). Alchemy and Artificial Intelligence (No. P-3244). Santa Monica, CA: Rand Corporation.

D'Souza, R. (2018). Symbolic AI v/s non-symbolic AI, and everything in between? Data Driven Investor, 19th October, 2018. Available at https://medium.com/datadriveninvestor/symbolic-ai-v-s-non-symbolic-ai-and-everything-in-between-ffcc2b03bc2e (last accessed June 24, 2019).

Dunbar, R. (1998). Grooming, Gossip, and the Evolution of Language. Cambridge, MA: Harvard University Press.

Durka, W., Babik, W., Ducroz, J.F., Heidecke, D., Rosell, F., Samjaa, R., Saveljev, A.P., Stubbe, A., Ulevi-ius, A. and Stubbe, M. (2005). Mitochondrial phylogeography of the Eurasian beaver *Castor fiber* L. Molecular Ecology 14(12): 3843-3856.

Durkheim, É. (1973). The dualism of human nature and its social conditions. *In*: Bellah, R.N. (ed.), Émile Durkheim on Morality and Society: Selected Writings. Chicago: Chicago University Press, pp. 149-163.

Dustin, J. (2018). Amazon scraps secret AI recruiting tool that showed bias against women. Reuters Business News 10th October, 2018. Available at https://www.reuters.com/article/us-amazon-com-jobs-automation-insight/amazon-scraps-secret-ai-recruiting-tool-that-showed-bias-against-women-idUSKCN1MK08G (last accessed June 24, 2019).

Dyall, S.D., Brown, M.T. and Johnson, P.J. (2004). Ancient invasions: From endosymbionts to organelles. Science 304(5668): 253-257.

Edelman, G.M. and Mountcastle, V.B. (1978). The Mindful Brain: Cortical Organization and the Group-Selective Theory of Higher Brain Function. Cambridge, MA: MIT Press.

Eisenstein, E. (2011). Divine Art, Infernal Machine: The Reception of Printing in the West from First Impressions to the Sense of an Ending. Philadelphia, PA: University of Pennsylvania Press.

Ela, J.-M. (1998). Western development has failed: Looking to a new Africa. Le Monde Diplomatique, October, 1998.

Elliot, L. (2015). OPEC bid to kill off US shale sends oil price down to 2009 low. The Guardian 7 December 2015. Available at https://www.theguardian.com/business/2015/dec/07/opec-plan-kill-us-shale-oil-price-down-seven-year-low (last accessed June 24, 2019).

Escobar, A. (2011). Sustainability: Design for the pluriverse. Development 54(2): 137-140.

Espressoenglish website (2019). Available at https://www.espressoenglish.net/1000-most-common-words-in-english/ (last accessed June 24, 2019).

European Commission (2010). Critical raw materials for the EU. Report of the Ad-hoc Working Group on defining critical raw materials. European Commission, June, 2010. Available at http://www.euromines.org/files/what-we-do/sustainable-development-issues/2010-report-critical-raw-materials-eu.pdf (last accessed June 24, 2019).

European Commission (2014). Critical raw materials for the EU. Report of the Ad-hoc Working Group on defining critical raw materials. European Commission, May, 2014. Available at https://eur-lex.europa.eu/legal-content/EN/TXT/?uri=CELEX:52014DC0297 (last accessed June 24, 2019).

European Parliament Committee on Legal Affairs (2016). Civil law rules on robotics (2015/2103 (INL)). Brussels, Belgium: European Parliament. Paragraph L. Retrieved from Available at http://www.europarl.europa.eu/sides/getDoc.do?pubRef=-//EP//NONSGML%2BCOMPARL%2BPE-582.443%2B01%2BDOC%2BPDF%2BV0//EN (last accessed June 24, 2019).

Evans, D. (2012). The Internet of Everything: How More Relevant and Valuable Connections Will Change the World. Cisco Internet Business Solutions Group (IBSG), Cisco Systems, Inc., San Jose, CA, USA, White Paper. [Online]. Available at https://www.cisco.com/c/dam/global/en_my/assets/ciscoinnovate/pdfs/IoE.pdf (last accessed June 24, 2019).

Ewing, J.A. (2017). Hollow ecology: Ecological modernization theory and the death of nature. Journal of World-Systems Research 23(1): 126-155.

Fedor, P., Va⊠hara, J., Havel, J., Malenovský, I. and Spellerberg, I. (2009). Artificial intelligence in pest insect monitoring. Systematic Entomology 34(2): 398-400. Available at https://onlinelibrary.wiley.com/doi/full/10.1111/j.1365-3113.2008.00461.x (last accessed June 24, 2019).

Feingold, A. (1994). Gender differences in personality: A meta-analysis. Psychological Bulletin 116: 429-456.

Feit, B., Gordon, C.E., Webb, J.K., Jessop, T.S., Laffan, S.W., Dempster, T. and Letnic, M. (2018). Invasive cane toads might initiate cascades of direct and indirect effects in a terrestrial ecosystem. Biological invasions 20: 1833-1847.

Feldman, M., Friedler, S.A., Moeller, J., Scheidegger, C. and Venkatasubramanian, S. (2015). Certifying and removing disparate impact. Proceedings of the 21st ACM SIGKDD International Conference on Knowledge Discovery and Data Mining, Sydney, Australia, August 10-13, 2015, pp. 259-268.

Feng, K., Davis, S.J., Sun, L. and Hubacek, K. (2015). Drivers of the US CO_2 emissions 1997–2013. Nature Communication 6: 7714.

Field, T., Healy, B., Goldstein, S., Perry, S., Bendell, D., Schanberg, S., Zimmerman, E.A. and Kuhn, C. (1988). Infants of depressed mothers show 'depressed' behavior even with nondepressed adults. Child Development 59: 1569-1579.

Firth, R. (1952). Ethical absolutism and the ideal observer. Philosophy and Phenomenological Research 12(March): 317-345.

Fischer, J., Dyball, R., Fazey, I., Gross, C., Dovers, S., Ehrlich, P.R., Brulle, R.J., Christensen, C. and Borden, R.J. (2012). Human behavior and sustainability. Frontiers in Ecology and the Environment 10(3): 153-160.

Fisher, J. (1952). The Fulmar. London: Collins.

Fisher-Kowalski, M. and Haberl, H. (1997). Tons, Joules, and money: Modes of production and their sustainability problems. Society and Natural Resources 10: 61-85.

Fiske, S.T. (2000). Stereotyping, prejudice, and discrimination at the seam between the centuries: Evolution, culture, mind, and brain. European Journal of Social Psychology 30: 299-322.

Forth, J., Wiggins, G.A. and McLean, A. (2010). Unifying conceptual spaces: Concept formation in musical creative systems. Minds and Machines 20(4): 503-532.

Fossat, P., Bacqué-Cazenave, J., De Deurwaerdère, P., Delbecque, J.P. and Cattaert, D. (2014). Anxiety-like behavior in crayfish is controlled by serotonin. Science 344(6189): 1293-1297.

Foster, E., Contestabile, M., Blazquez, J., Manzano, B., Workman, M. and Shah, N. (2017). The unstudied barriers to widespread renewable energy deployment: Fossil fuel price responses. Energy Policy 103: 258-264.

Frappier, V. and Najmanovich, R. (2015). Vibrational entropy differences between mesophile and thermophile proteins and their use in protein engineering. Protein Science 24(4): 474-483.

Frey, C.B. and Osborne, M.A. (2017). The future of employment: How susceptible are jobs to computerisation? Technological Forecasting and Social Change 114: 254-280.

Friesdorf, R., Conway, P. and Gawronski, B. (2015). Gender differences in responses to moral dilemmas: A process dissociation analysis. Personality and Social Psychology Bulletin 41(5): 696-713.

Fujii, H. and Managi, S. (2013). Which industry is greener? An empirical study of nine industries in OECD countries. Energy Policy 57: 381-388.

Fukuda-Parr, S. (2010). Reducing inequality – The missing MDG: A content review of PRSPs and bilateral donor policy statements. IDS Bulletin-Institute of Development Studies 41(1): 26-35.

Fumagalli, M., Ferrucci, R., Mameli, F., Marceglia, S., Mrakic- Sposta, S., Zago, S., Lucchiari, C., Consonni, D., Nordio, F., Pravettoni, G. and Cappa, S. (2010). Gender-related differences in moral judgments. Cognitive Processing 11: 219-226.

Funtowicz, S. and Ravetz, J.R. (1994). Emergent complex systems. Futures 26(6): 568-582.

G20 (2017). G20 Leaders' Declaration: Shaping an Interconnected World. G20 Germany 2017 meetings, Hamburg, July 7-8. Available at http://www.g20.utoronto.ca/2017/2017-G20-leaders-declaration.html (last accessed June 24, 2019).

Gallant, A., Sadinski, W., Brown, J., Senay, G. and Roth, M. (2018). Challenges in complementing data from ground-based sensors with satellite-derived products to measure ecological changes in relation to climate – lessons from temperate wetland-upland landscapes. Sensors 18(3): 880.

Gärdenfors, P. (2000). Conceptual Spaces. Cambridge, MA: MIT Press.

Gardner, H. (1983). Frames of Mind: The Theory of Multiple Intelligences. New York: Basic Books.

Gaston, K.J. and O'Neill, M.A. (2004). Automated species identification: Why not? Philosophical Transactions of the Royal Society of London. Series B: Biological Sciences 359(1444): 655-667.

Geddes, P. (1915). Cities in Evolution. London: Williams.

Gelter, H. (2000). Friluftsliv: The Scandinavian philosophy of outdoor life. Canadian Journal of Environmental Education 5(1): 77-92.

GEO BON (2019). GEO BON's 2025 Vision Statement and Goals. Available at http://www.unoosa.org/oosa/en/spaceobjectregister/index.html (last accessed June 24, 2019).

Geraci, R.M. (2008). Apocalyptic AI: Religion and the promise of artificial intelligence. Journal of the American Academy of Religion 76(1): 138-166.

Gifford, R. (2011). The dragons of inaction: Psychological barriers that limit climate change mitigation and adaptation. American Psychologist 66(4): 290 302. Available at https://www.researchgate.net/profile/Robert_Gifford3/publication/254734365_The_Dragons_of_Inaction_Psychological_Barriers_That_Limit_Climate_Change_Mitigation_and_Adaptation/links/0c96052047aaad383e000000/The-Dragons-of-Inaction-Psychological-Barriers-That-Limit-Climate-Change-Mitigation-and-Adaptation.pdf (last accessed June 24, 2019).

Gilligan, C. (1982). In a Different Voice: Psychological Theory and Women's Development. Cambridge, MA: Harvard University Press.

Glacken, C.J. (1976). Traces on the Rhodian Shore: Nature and Culture in Western Thought from Ancient Times to the End of the Eighteenth Century. Berkeley, CA: University of California Press.

Goldsmith, E. (1988). The Great U-turn: De-industrializing Society. Green Books, London.

Gómez Gutiérrez, E. (2019). Women in Artificial Intelligence: Mitigating the Gender Bias. European Commission: HUMAINT. Available at https://ec.europa.eu/jrc/communities/en/community/humaint/news/women-artificial-intelligence-mitigating-gender-bias (last accessed June 24, 2019).

Gould, S.J. (1990). Wonderful Life: The Burgess Shale and the Nature of History. New York: WW Norton and Company.

Grace, K., Salvatier, J., Dafoe, A., Zhang, B. and Evans, O. (2017). When will AI exceed human performance. Evidence from AI experts. Opensource, 20. Available at https://arxiv.org/pdf/1705.08807.pdf (last accessed June 24, 2019).

Graef, A. (2014). Elon Musk: We Are "Summoning a Demon" with Artificial Intelligence, UPI (Oct. 27, 2014, 7:50 AM). Available at https://www.quora.com/Elon-Musk-said-in-an-interview-It%E2%80%99s-like-we-are-summoning-a-demon-when-talking-about-artificial-intelligence-What-are-your-thoughts-on-that (last accessed June 24, 2019).

Graff, J.L. (2006). The battle to save the cave. Time Magazine, 16th June, 2006.

Green, W.M. (1942). The dying world of Lucretius. The American Journal of Philology 63(1): 51-60.

Greenstone, M. and Fan, C.Q. (2018). Introducing the Air Quality Life Index. University of Chicago Energy Policy Institute. Available at https://aqli.epic.

uchicago.edu/wp-content/uploads/2018/11/AQLI-Report.111918-2.pdf (last accessed June 24, 2019).

Grin, J., Rotmans, J. and Schot, J. (2010). Transitions to Sustainable Development. London: Routledge.

Grinbaum, A., Chatila, R., Devillers, L., Ganascia, J.-B., Tessier, C. and Dauchet, M. (2017). Ethics in robotics research. CERNA mission and context. IEEE Robotics and Automation Magazine 24(3): 139-145.

Grossman, G.M. and Krueger, A.B. (1995). Economic growth and the environment. Quarterly Journal of Economics 110: 353-377.

Gruntman, M. and Novoplansky, A. (2004). Physiologically mediated self/nonself discrimination in roots. Proceedings of the National Academy of Sciences USA 101: 3863-3867.

Gudynas, E. (2011). BuenVivir: Today's tomorrow. Development 54: 441-447.

Häder, D.P. (1997). Penetration and effects of solar UV-B on phytoplankton and macroalgae. *In*: Rozema, J., Gieskes, W.W.C., an de Geijn, S.C., Nolan, C. and de Boois, H. (eds.), UV-B and Biosphere. Dordrecht: Springer, pp. 4-13.

Haldane, J.B. (1926). On being the right size. Harper's Magazine 152: 424-427.

Hallowell, N., Amoore, L., Caney, S. and Waggett, P. (2019). Ethical issues arising from the police use of live facial recognition technology. Interim report of the Biometrics and Forensics Ethics Group Facial Recognition Working Group, February 2019. Available at https://assets.publishing.service.gov.uk/government/uploads/system/uploads/attachment_data/file/781745/Facial_Recognition_Briefing_BFEG_February_2019.pdf (last accessed June 24, 2019).

Halopka, R. (2017). The high cost of soil erosion. Farm Progress, 27th September, 2017. Available at https://www.farmprogress.com/soil-health/high-cost-soil-erosion (last accessed June 24, 2019).

Hamaguchi, N. and Kondo, K. (2018). Regional Employment and Artificial Intelligence in Japan. Research Institute of Economy, Trade and Industry (RIETI).

Hancock, L., Ralph, N. and Ali, S.H. (2017). Bolivia's lithium frontier: Can public private partnerships deliver a minerals boom for sustainable development? Journal of Cleaner Production 178: 551-560.

Harari, Y.N. (2015). Sapiens: A Brief History of Humankind. New York: Harper.

Harari, Y.N. (2017). Homo Deus: A Brief History of Tomorrow. New York: Harper.

Haraway, D. (2000). A cyborg manifesto: Science, technology, and socialist-feminism in the late 20th century. *In*: Bell, D. and Kennedy, B.M. (eds.), The Cybercultures Reader. London: Routledge, pp. 291-324. Available at http://faculty.georgetown.edu/irvinem/theory/Haraway-CyborgManifesto-1.pdf (last accessed June 24, 2019).

Hardin, G. (1968). The tragedy of the commons. Science 162(3859): 1243-1248.

Harris, M. (1999). Lament for an Ocean: The Collapse of the Atlantic Cod Fishery: A True Crime Story. Toronto: McClelland and Stewart Inc.

Harrisson, T.H. and Buchan, J.N. (1934). A field study of the St Kilda wren (*Troglodytes troglodyteshirtensis*), with especial reference to its numbers, territory and food habits. Journal of Animal Ecology 3(2): 133-145.

Hayes, A.W. (2005). The precautionary principle. Arhiv za Higijenu Rada iToksikologiju 56(2): 161-166. Available at https://www.researchgate.net/publication/7774245_The_Precautionary_Principle (last accessed June 24, 2019).

He, F., Fromion, V. and Westerhoff, H.V. (2013). (Im) Perfect robustness and adaptation of metabolic networks subject to metabolic and gene-expression regulation: Marrying control engineering with metabolic control analysis. BMC

Systems Biology 7: 131. Available at https://bmcsystbiol.biomedcentral.com/articles/10.1186/1752-0509-7-131 (last accessed June 24, 2019).

Helbing, D., Frey, B.S., Gigerenzer, G., Hafen, E., Hagner, M., Hofstetter, Y., van den Hoven, J., Zicari, R.V. and Zwitter, A. (2016). Big Data revolution: Behavioural control or digital democracy? It's time to decide. Downloaded from: Available at https://www.researchgate.net/profile/Dirk_Helbing/publication/303813069_Behavioural_Control_or_Digital_Democracy_-_A_Digital_Manifesto/links/5755309208ae10d9337a47a2.pdf (last accessed June 24, 2019).

Helbing, D., Frey, B.S., Gigerenzer, G., Hafen, E., Hagner, M., Hofstetter, Y., van den Hoven, J., Zicari, R.V. and Zwitter, A. (2017). Will democracy survive big data and artificial intelligence? Scientific American, February 25, 2017. Available at https://www.scientificamerican.com/article/will-democracy-survive-big-data-and-artificial-intelligence/ (last accessed June 24, 2019).

Hennink, M., Kiiti, N., Pillinger, M. and Jayakaran, R. (2012). Defining empowerment: Perspectives from international development organisations. Development in Practice 22: 202-215.

Heylighen, F. (2016). Stigmergy as a universal coordination mechanism I: Definition and components. Cognitive Systems Research 38: 4-13.

Hickes, F. (Translator) (1894) Lucian's True History. A.H. Bullen, 18 Cecil Court, London.

Hikmany, A.N., Kader, A., Zubaidah, S. and Othman, A. (2015). Legal issues of land acquisition in Zanzibar. International Journal of Business Management and Economic Studies: Available at https://www.researchgate.net/profile/Abdul_Nasser_Hikmany/publication/304395039_Legal_Issues_of_Land_Acquisition_in_Zanzibar/links/576e23b208ae10de6395d88e/Legal-Issues-of-Land-Acquisition-in-Zanzibar.pdf (last accessed June 24, 2019).

Hillel, D. (1991). Out of the Earth: Civilization and the Life of the Soil. Berkeley, California: University of California Press.

Hirzel, A.H., Helfer, V. and Metral, F. (2001). Assessing habitat-suitability models with a virtual species. Ecological Modelling 145(2-3): 111-121.

Hoggan, J. (2009). Climate Cover-up: The Crusade to Deny Global Warming. Vancouver: Greystone Books.

Holley, P. (2015). Apple co-founder on artificial intelligence: The future is scary and very bad for people. Washington Post. 24th March, 2015. Available at http://www.washingtonpost.com/blogs/the-switch/wp/2015/03/24/apple-cofounder-on-artificial-intelligence-the-future-is-scary-and-very-bad-for-people/ (last accessed June 24, 2019).

Horace (2002). Odes. New York: Random House, Inc.

Horowitz, M.C. (2018). Artificial intelligence, international competition, and the balance of power (May 2018). Texas National Security Review.

Hughes, L., Steffen, W., Alexander, D. and Rice, M. (2017). Climate Change: A Deadly Threat to Coral Reefs. Climate Council of Australia. Available at https://uploads.guim.co.uk/2017/04/11/CC_Report_1.pdf (last accessed June 24, 2019).

Humphreys Bebbington, D. (2013). Extraction, inequality and indigenous people: Insights from Bolivia. Environmental Science and Policy 33: 438-446.

Hunt, J. (2015). Headline News – Today, 29 August 1930. St Kilda Club. 28th August, 2015. Available at https://www.stkildaclub.co.uk/headline-news-today-29-august-1930 (last accessed June 24, 2019).

Hunt, E. (2016). Tay, Microsoft's AI chatbot, gets a crash course in racism from Twitter. The Guardian, 24th March, 2016. Available at https://www.theguardian.com/

technology/2016/mar/24/tay-microsofts-ai-chatbot-gets-a-crash-course-in-racism-from-twitter (last accessed June 24, 2019).

Husband, C. (1995). The morally active practitioner and the ethics of anti-racist social work. *In*: Hugman, R. and Smith, D. (eds.), Ethical Issues in Social Work. London: Routledge, pp. 84-103.

Hyde, J.S. (2016). Sex and cognition: Gender and cognitive functions. Current Opinion in Neurobiology 38: 53-56.

Ichihara, M. and Harding, A. (1995). Human rights, the environment and radioactive waste: A study of the Asian Rare Earth case in Malaysia. Review of European Community and International Environmental Law 4(1): 1-14.

Ingold, D. and Soper, S. (2016). Amazon doesn't consider the race of its customers. Should it? Bloomberg 21st April, 2016. Available at https://www.bloomberg.com/graphics/2016-amazon-same-day/ (last accessed June 24, 2019).

IPBES (2019). IPBES Global Assessment Summary for Policymakers. Available at https://www.ipbes.net/sites/default/files/downloads/spm_unedited_advance_for_posting_htn.pdf (last accessed June 24, 2019).

Isaac, A. (2018). Bank of England deputy warns UK economy entering 'menopausal' phase. The Telegraph, Business Section, 16th May, 2018. Available at https://www.telegraph.co.uk/business/2018/05/15/menopausal-uk-economy-risks-once-in-a-century-slump-warns-deputy/ (last accessed June 24, 2019).

Ismail, N. (2017). Revenue for cognitive/AI systems to top $47 billion by 2020. [Online] Available at: http://www.information-age.com/revenue-ai-systems-top-47-billion-2020-123465508/ (last accessed June 24, 2019).

Ivanov, S. and Webster, C. (2017). The robot as a consumer: A research agenda. Paper presented at the 'Marketing: Experience and Perspectives' Conference, 29-30 June 2017, University of Economics-Varna, Bulgaria. Available at https://www.researchgate.net/profile/Stanislav_Ivanov/publication/316587652_THE_ROBOT_AS_A_CONSUMER_A_RESEARCH_AGENDA/links/5905962fa6fdccd580d2420e/THE-ROBOT-AS-A-CONSUMER-A-RESEARCH-AGENDA.pdf (last accessed June 24, 2019).

Jackson, T. (2009). Prosperity without growth? The transition to a sustainable economy. London: The Sustainable Development Commission. Available at http://www.sd-commission.org.uk/data/files/publications/prosperity_without_growth_report.pdf (last accessed June 24, 2019).

Jacob, E.B., Becker, I., Shapira, Y. and Levine, H. (2004). Bacterial linguistic communication and social intelligence. Trends in Microbiology 12(8): 366-372. Available at http://www.israela-becker.me/docs/ben-jacob_becker_shapira_levine_2004.pdf (last accessed June 24, 2019).

Jaffee, S. and Hyde, J.S. (2000). Gender differences in moral orientation: A meta-analysis. Psychological Bulletin 126: 703-726.

Jansen, M.A., Gaba, V. and Greenberg, B.M. (1998). Higher plants and UV-B radiation: Balancing damage, repair and acclimation. Trends in Plant Science 3(4): 131-135.

Jepson, P. and Ladle, R.J. (2015). Nature apps: Waiting for the revolution. Ambio 44(8): 827-832.

Jevons, W.S. (2001). Of the economy of fuel. Organization and Environment 14: 99-104.

Jiabao, L. and Jie, L. (2009). Rare earth industry adjusts to slow market. China Daily [online], available at http://www.chinadaily.com.cn/bw/2009-09/07/content_8660849.htm (last accessed June 24, 2019).

Johnson, D.G. and Wetmore, J.M. (2009). Technology and Society Building our Sociotechnical Future. Cambridge, MA: The MIT Press.

Johnson, S.S., Hebsgaard, M.B., Christensen, T.R., Mastepanov, M., Nielsen, R., Munch, K., Brand, T., Gilbert, M.T.P., Zuber, M.T., Bunce, M. and Rønn, R. (2007). Ancient bacteria show evidence of DNA repair. Proceedings of the National Academy of Sciences USA 104(36): 14401-14405.

Jorgenson, A. and Clark, B. (2011). Societies consuming nature: A panel study of the ecological footprints of nations, 1960–2003. Social Science Research 40: 226-244.

Jorgenson, A. and Rice, J. (2012). The sociology of ecologically unequal exchange in comparative perspective. *In*: Babones, S. and Chase-Dunn, C. (eds.), Handbook of World-Systems Analysis: Theory and Research. New York: Routledge Press.

Kant, I. (2002). Groundwork of the Metaphysics of Morals (translated by Wood). A.R. New Haven, CT: Yale University Press.

Karban, R., Huntzinger, M. and McCall, A.C. (2004). The specificity of eavesdropping on sagebrush by other plants. Ecology 85: 1846-1852.

Kaufman, A.S. (2018). Forward. *In*: Flanagan, D.P. and McDonough, E.M. (eds.), Contemporary Intellectual Assessment: Theories, Tests, and Issues. Fourth Edition. New York: Guilford Publications, p. ix.

Kaya, Y., Kayci, L., Tekin, R. and Faruk Ertuğrul, Ö. (2014). Evaluation of texture features for automatic detecting butterfly species using extreme learning machine. Journal of Experimental and Theoretical Artificial Intelligence 26(2): 267-281.

Kelley, T.D. and Long, L.N. (2010). Deep Blue cannot play checkers: The need for generalized intelligence for mobile robots. Journal of Robotics, 2010. Available at https://www.hindawi.com/journals/jr/2010/523757/ (last accessed June 24, 2019).

Kelly, K. (2012). Better than human: Why robots will—and must—take our jobs. Wired. Available at https://www.wired.com/2012/12/ff-robots-will-take-our-jobs/ (last accessed June 24, 2019).

Kelly, L.W., Barott, K.L., Dinsdale, E., Friedlander, A.M., Nosrat, B., Obura, D., Sala, E., Sandin, S.A., Smith, J.E., Vermeij, M.J. and Williams, G.J. (2012). Black reefs: Iron-induced phase shifts on coral reefs. The ISME Journal 6(3): 638-649.

Kelly, B. and Resnick, P. (2014). A single IQ score over 70 supports finding of no intellectual disability, despite conflicting test results and expert testimony. Journal of the American Society of Psychiatry Law 42: 515-517.

Kenya National Bureau of Statistics (2010). 2009 Kenya population and housing census: Population and household distribution by socio-economic characteristics, Vol 2. Nairobi: KNBS.

Kitano, H. (2002). Systems biology: A brief overview. Science 295(5560): 1662-1664.

Klare, F.B., Burge, M.J., Klontz, J.C., Vorder Bruegge, R.W. and Jain, A.K. (2012). Face recognition performance: Role of demographic information. IEEE Transactions on Information Forensics and Security 7(6): 1789-1801.

Kollmuss, A. and Agyeman, J. (2002). Mind the gap: Why do people act environmentally and what are the barriers to pro-environmental behavior? Environmental Education Research 8(3): 239-260.

Korinek, A. and Stiglitz, J.E. (2017). Artificial intelligence and its implications for income distribution and unemployment (No. w24174). National Bureau of Economic Research.

Kothari, A., Demaria, F. and Acosta, A. (2014). Buenvivir, degrowth and ecological Swaraj: Alternatives to sustainable development and the green economy. Development 57(3-4): 362-375.

Koulouri, A. and Moccia, J. (2014). Saving water with wind energy. The European Wind Energy Association (EWEA). Available at http://www.ewea.org/fileadmin/files/library/publications/reports/Saving_water_with_wind_energy.pdf (last accessed June 24, 2019).

KPMG International (2016). Rise of the Humans: The Integration of Digital and Human Labor. KPMG International Cooperative, November. Available at https://assets.kpmg.com/content/dam/kpmg/xx/pdf/2016/11/rise-of-the-humans.pdf (last accessed June 24, 2019).

Kramer, J. and Meunier, J. (2016). Kin and multilevel selection in social evolution: A never-ending controversy? F1000 Research, 5. Available at https://www.ncbi.nlm.nih.gov/pmc/articles/PMC4850877/ (last accessed June 24, 2019).

Kumar, A., Khorwal, R. and Chaudhary, S. (2016). A Survey on sentiment analysis using swarm intelligence. Indian Journal of Science and Technology 9(39): 1-7. Available at http://www.indjst.org/index.php/indjst/article/viewFile/100766/74445 (last accessed June 24, 2019).

Kurzweil, R. (2005). The Singularity is Near: When Humans Transcend Biology. New York: Viking.

Kuznets, S. (1955). Economic growth and income inequality. American Economic Review 45(1): 1-28.

LaChat, M.R. (1986). Artificial intelligence and ethics: An exercise in the moral imagination. AI Magazine 7(2): 70-79.

Landy, F.J. (2006). The long, frustrating, and fruitless search for social intelligence: A cautionary tale. *In*: Murphy, K.R. (ed.), A Critique of Emotional Intelligence: What Are the Problems and How Can They Be fixed? Mahwah, NJ: Lawrence Erlbaum Associates, Publishers, pp. 81-123.

Langford, M. (2010). A poverty of rights: Six ways to fix the MDGs. IDS Bulletin-Institute of Development Studies 41(1): 83-91.

Latour, B. (2005). Reassembling the Social: An Introduction to Actor-Network-Theory. Oxford: Oxford University Press.

Law, J. (1987). Technology and heterogeneous engineering: The case of the Portuguese expansion. *In*: Bijker, W., Hughes, T. and Pinch, T. (eds.), The Social Construction of Technological Systems. Cambridge, MA: MIT Press, pp. 111-134.

Lazer, D., Kennedy, R., King, G. and Vespignani, A. (2014). Google flu trends still appears sick: An evaluation of the 2013–2014 flu season. 13th March, 2014. Rochester, N.Y.: Social Science Electronic Publishing, Inc.

Lee, H. (2014). Paging Dr. Watson: IBM's Watson supercomputer now being used in healthcare. Journal of AHIMA 85(5): 44-47.

Legg, S. and Hutter, M. (2007). A collection of definitions of intelligence. Frontiers in Artificial Intelligence and Applications 157: 17-24. Available at https://arxiv.org/pdf/0706.3639.pdf%20a%20collection%20of%20definitions%20of%20intelligence (last accessed June 24, 2019).

Legg, C.J. and Nagy, L. (2006). Why most conservation monitoring is, but need not be, a waste of time. Journal of Environmental Management 78(2): 194-199.

Lehmann, E. (1934). Biologischer Wille. Wege und Zielebiologischer Arbeit imneuen Reich [Biological Will: Means and Goals of Biological Work in the New Reich]. München, 1934, pp. 10-11.

Lei, X., Hu, Y., McArdle, J.J., Smith, J.P. and Zhao, Y. (2012). Gender differences in cognition among older adults in China. Journal of Human Resources 47(4): 951-971.

Levy, J. (2009). Tin. New York: The Rosen Publishing Group, Inc.

Levy, F. (2018). Computers and populism: Artificial intelligence, jobs, and politics in the near term. Oxford Review of Economic Policy 34(3): 393-417.

Lewis, H.W. (1990). Technological Risk. New York: W.W. Norton, p. 113.

Lewis, J.A. (2016). Managing risk for the internet of things. Washington, DC: Center for Strategic and International Studies. Available at https://cybersummit.info/sites/cybersummit.info/files/Excerpt_Managing_Risk_for_the_IoT.pdf (last accessed June 24, 2019).

Li, D. and Du, Y. (2017). Artificial Intelligence with Uncertainty. Boca Raton, FL: CRC Press.

Li, X., Chen, Z., Chen, Z. and Zhang, Y. (2013). A human health risk assessment of rare earth elements in soil and vegetables from a mining area in Fujian Province, Southeast China. Chemosphere 93: 1240–1246.

Lighthill Debate (1973). Available at https://www.youtube.com/watch?v=03p2CADwGF8 (last accessed June 24, 2019).

Lisbon Treaty (2009). The Lisbon Treaty. Available at http://www.europarl.europa.eu/about-parliament/en/powers-and-procedures/the-lisbon-treaty (last accessed June 24, 2019).

Lister, K.N., Lamare, M.D. and Burritt, D.J. (2010). Sea ice protects the embryos of the Antarctic sea urchin, *Sterechinusneumayeri*, from oxidative damage due to naturally enhanced levels of UV-B radiation. Journal of Experimental Biology 213(11): 1967-1975.

Lockwood, M. (2010). The economics of personal carbon trading. Climate Policy 10: 447-461.

Loorbach, D., van Bakel, J.C., Loorbach, D.A., Whiteman, G.M. and Rotmans, J. (2007). Business strategies for transitions towards sustainable Systems. (No. ERS-2007-94-ORG). Erasmus Research Institute of Management (ERIM).

Lorde, A. (1984). Sister Outsider. Freedom, CA: The Crossing Press.

Lorimer, H. (2002). Sites of authenticity: Scotland's new parliament and official representations of the nation. *In*: Harvey, D., Jones, R., McInroy, N. and Milligan, C. (eds.), Celtic Geographies: Old Cultures, New Times. London, Routledge, pp. 91-108.

Lovelock, J. (1988). The Earth as a living organism. *In*: Wilson E.O. and Peter, F.M. (eds.), Biodiversity. Washington DC: National Academies Press, pp. 486-489.

Lucidi, P.B. and Nardi, D. (2018). Companion robots: The hallucinatory danger of human-robot interactions. Available at http://www.aies-conference.com/2018/contents/papers/main/AIES_2018_paper_60.pdf (last accessed June 24, 2019).

Lyons, M. (2007). Place-shaping: A Shared Ambition for the Future of Local Government. Executive summary. London, The Stationary Office.

Macaulay, K. (1764). The History of St Kilda. London: Becket and De Hondt.

Maccoby, E.E. and Jacklin, C.N. (1974). The Psychology of Sex Differences. Stanford, CA: Stanford University Press.

MacDonald, F. (2001). St Kilda and the sublime. Ecumene 8(2): 151-174.

Mace, F.C., Hock, M.L., Lalli, J.S., West, B.J., Belfiore, P., Pinter, E. and Brown, D.K. (1988). Behavioral momentum in the treatment of noncompliance. Journal of Applied Behavior Analysis 21(2): 123-141.

MacIntyre, A. (1999). Dependent Rational Animals: Why Human Beings Need the Virtues. Chicago and La Salle, Illinois: Open Court.

Mack, E. (2015). Bill Gates says you should worry about artificial intelligence. Available at https://www.forbes.com/sites/ericmack/2015/01/28/bill-gates-also-worries-artificial-intelligence-is-a-threat/#3862337a651f (last accessed June 24, 2019).

MacKenzie, D. (2018). 'Making', 'taking' and the material political economy of algorithmic trading. Economy and Society 47(4): 501-523.

Mackintosh, N.J. (1998). IQ and Human Intelligence. New York: Oxford University Press.

MacLean, L. (1838). Sketches on the Island of St Kilda. Glasgow: W.R. McPhun.

Madakam, S. (2015). Internet of things: Smart things. International Journal of Future Computer and Communication 4(4): 250-253.

Madrigal, A.C. (2013). IBM's Watson memorized the entire 'urban dictionary,' then his overlords had to delete it. Atlantic, January 10, 2013. Available at http://www.theatlantic.com/technology/archive/2013/01/ibms-watson-memorized-the-entire-urban-dictionary-then-his-overlords-had-to-delete-it/267047/ (last accessed June 24, 2019).

Mandeville, B. (1705). The Grumbling Hive: or, Knaves Turn'd Honest. Available at https://andromeda.rutgers.edu/~jlynch/Texts/hive.html (last accessed June 24, 2019).

Manyika, J., Lund, S., Chui, M., Bughin, J., Woetzel, J., Batra, P., Ko, R. and Sanghvi, S. (2017). What the future of work will mean for jobs, skills, and wages. McKinsey Quarterly, November. 2017. Available at https://www.mckinsey.com/global-themes/future-of-organizations-and-work/what-the-future-of-work-will-mean-for-jobs-skills-and-wages (last accessed June 24, 2019).

Margolis, J. and Fisher, A. (2003). Unlocking the Clubhouse: Women in Computing. Boston, MA: MIT Press.

Markel, R.W. and Shurin, J.B. (2015). Indirect effects of sea otters on rockfish (*Sebastes* spp.) in giant kelp forests. Ecology 96(11): 2877-2890.

Marlowe, H.A. (1986). Social intelligence: Evidence for multidimensionality and construct independence. Journal of Educational Psychology 78(1): 52-58.

Martin, M. (1698). A Late Voyage to St Kilda. London: D. Brown and D. Goodwin. Available at https://www.undiscoveredscotland.co.uk/usebooks/martin-stkilda/chapter04.html (last accessed June 24, 2019).

Masch, V.A. (2017). Shifting the paradigm in superintelligence. Review of Economics and Finance 8: 17-30.

Mathews, F.D. (2014). The Ecology of Democracy: Finding Ways to Have a Stronger Hand in Shaping Our Future. Dayton, OH: Kettering Foundation Press.

Mauser, W., Klepper, G., Rice, M., Schmalzbauer, B.S., Hackmann, H., Leemans, R. and Moore, H. (2013). Transdisciplinary global change research: The co-creation of knowledge for sustainability. Current Opinion in Environmental Sustainability 5(3-4): 420-431.

Mayer, E.A., Padua, D. and Tillisch, K. (2014). Altered brain-gut axis in autism: Comorbidity or causative mechanisms? BioEssays 36: 933-939.

Mbiti, J.S. (1969). African Religions and Philosophy. London: Heinemann.

McCarthy, J., Minsky, M.L., Rochester, N. and Shannon, C.E. (1955). Proposal for the 1956 Dartmouth Summer Research Project on Artificial Intelligence. Dartmouth College, Hanover, NH, USA. Available at https://fermatslibrary.com/s/a-

proposal-for-the-dartmouth-summer-research-project-on-artificial-intelligence (last accessed June 24, 2019).
McLuhan, M., Fiore, Q. and Agel, J. (1967). The Medium is the Massage: An Inventory of Effects. New York: Bantam Books.
McNeill, J.R. (2001). Something New Under the Sun: An Environmental History of the Twentieth-Century World. New York: Norton.
Mead, G.H. (1934). Mind, Self and Society. Chicago, University of Chicago Press.
Meekosha, H. and Shuttleworth, R. (2009). What's so 'critical' about critical disability studies? Australian Journal of Human Rights 15: 47-75.
Meier, P. (2014). Big data and next generation humanitarian technologies. Available at https://www.youtube.com/watch?v=FPHPi0LYz7o&feature=youtu.be (last accessed June 24, 2019).
Mekonnen, M.M. and Hoekstra, A.Y. (2010). The green, blue and grey water footprint of crops and derived crop products. Value of Water Research Report Series No. 47. UNESCO-IHE, Delft, the Netherlands. Water footprint scenarios for 2050, 49.
Mekonnen, M.M. and Hoekstra, A.Y. (2011). The green, blue and grey water footprint of crops and derived crop products. Hydrology and Earth System Sciences 15(5): 1577-1600.
Mekonnen, M.M. and Hoekstra, A.Y. (2012). A global assessment of the water footprint of farm animal products. Ecosystems 15(3): 401-415.
Microsoft (2018). The Future Computed: Artificial Intelligence and its Role in Society. Redmond, WA: Microsoft. Available at https://blogs.microsoft.com/uploads/2018/02/The-Future-Computed_2.8.18.pdf (last accessed June 24, 2019).
Mikolov, T., Joulin, A. and Baroni, M. (2016). A roadmap towards machine intelligence. *In*: Gelbukh, A. (ed.), International Conference on Intelligent Text Processing and Computational Linguistics. Cham: Springer, pp. 29-61.
Milanovic, B., Lindert, P.H. and Williamson, I.G. (2007). Measuring Ancient Equality. Munich Personal RePEc Archive, October 2007.
Millen, J.V., Irwin, A. and Kim, J.Y. (2000). Introduction: What is growing? Who is dying? *In*: Kim. J.Y., Millen, J.V., Irwin, A. and Gershman, J. (eds.), Dying for Growth: Global Inequality and the Health of the Poor. Monroe, ME: Common Courage Press, pp. 3-10.
Mill, J.S. (1843). A System of Logic, Ratiocinative and Inductive: Being a Connected View of the Principles of Evidence and the Methods of Scientific Investigation. London: John W. Parker.
Miotto, R., Li, L., Kidd, B.A. and Dudley, J.T. (2016). Deep patient: An unsupervised representation to predict the future of patients from the electronic health records. Scientific Reports, 6, 26094. Available at https://www.nature.com/articles/srep26094 (last accessed June 24, 2019).
Mitchell, S., Villa, N., Stewart-Weeks, M. and Lange, A. (2013). The Internet of Everything for Cities: Connecting People, Process, Data, and Things to Improve the 'Livability' of Cities and Communities. Cisco Internet Business Solutions Group (IBSG), Cisco Systems, Inc., San Jose, CA, USA, White Paper. Available at http://www.cisco.com/web/strategy/docs/gov/everything-for-cities.pdf (last accessed June 24, 2019).
Moilanen, J. (2012). Emerging hackerspaces–peer production generation. *In*: Hammouda, I., Lundell, B., Mikkonen, T. and Scacch, W. (eds.), i*IFIP International Conference on Open Source Systems*. New York: Springer, pp. 94-111.

Mol, A. (1995). The Refinement of Production: Ecological Modernization Theory and the Chemical Industry. Utrecht, The Netherlands: Van Arkel.

Montzka, S.A., Dutton, G.S., Yu, P., Ray, E., Portmann, R.W., Daniel, J.S., Kuijpers, L., Hall, B.D., Mondeel, D., Siso, C. and Nance, J.D. (2018). An unexpected and persistent increase in global emissions of ozone-depleting CFC-11. Nature 557(7705): 413.

Moran, S., Pantidi, N., Rodden, T., Chamberlain, A., Griffiths, C., Zilli, D., Merrett, G. and Rogers, A. (2014). Listening to the forest and its curators: Lessons learnt from a bioacoustic smartphone application deployment. *In*: Jones, M., Palanque, P., Schmidt, A. and Grossman, T. (eds.), Proceedings of the 32nd Annual ACM Conference on Human Factors in Computing Systems. Toronto: ACM, pp. 2387-2396. Available at http://eprints.nottingham.ac.uk/37417/1/paper399.pdf (last accessed June 24, 2019).

Moravec, H. (1988). Mind Children: The Future of Robot and Human Intelligence. Cambridge, MA: Harvard University Press.

Morgan, C.L. (1894). An Introduction to Comparative Psychology. London: The Walter Scott Publishing Company.

Morgan, S. (2017). Cybercrime damages $6 trillion by 2021. Cybersecurity Ventures. 16th October, 2017. Available at https://cybersecurityventures.com/hackerpocalypse-cybercrime-report-2016/ (last accessed June 24, 2019).

Muller-Karger, F.E., Hestir, E., Ade, C., Turpie, K., Roberts, D.A., Siegel, D., Miller, R.J., Humm, D., Izenberg, N., Keller, M. and Morgan, F. (2018). Satellite sensor requirements for monitoring essential biodiversity variables of coastal ecosystems. Ecological Applications 28(3): 749-760.

Murray, A., Skene, K. and Haynes, K. (2017). The circular economy: An interdisciplinary exploration of the concept and application in a global context. Journal of Business Ethics 140(3): 369-380. Available at https://static1.squarespace.com/static/5a5f0a586f4ca3fbe272618d/t/5aec63a31ae6cff5ed8ddad1/1525441444970/circular+economy.pdf (last accessed June 24, 2019).

National Records of Scotland (1930). Petition to Secretary of State for Scotland signed by the islanders of St Kilda requesting government assistance to leave the island, 10 May 1930. National Records of Scotland, Agriculture and Fisheries files, AF57/26/3 Available at https://www.nrscotland.gov.uk/files//research/StKildaFeature-AF57-26-3Transcription.pdf (last accessed June 24, 2019).

Neumayer, E. (2002). Do democracies exhibit stronger international environmental commitment? A cross-country analysis. Journal of Peace Research 39(2): 139-164.

New York Times (1984). U.S. weather agency held liable in storm deaths. December 22, 1984. Available at https://www.nytimes.com/1984/12/22/us/us-weather-agency-held-liable-in-storm-deaths.html (last accessed June 24, 2019).

Nietzsche, F. (1986). Human all too Human, Volume 1. Translated by R.J. Holland. Cambridge: Cambridge University Press.

Nietzsche, F. (1990). Twilight of the Idols: The Antichrist. Translated by R.J. Hollingdale. New York: Penguin.

Nordhaus, W.D. and Tobin, J. (1972). Is growth obsolete? *In*: Nordhaus, W.D. and Tobin, J. (eds.), Economic Research: Retrospect and Prospect, Volume 5. Economic growth. Cambridge, MA: The National Bureau of Economic Research, pp. 1-80. Available at https://www.nber.org/chapters/c7620.pdf (last accessed June 24, 2019).

NSTC (2016). Preparing for the Future of Artificial Intelligence. Executive Office of the President. National Science and Technology Council Committee on Technology.

US Government. October, 2016. Available at https://www.hartnell.edu/sites/default/files/library_documents/preparing_for_the_future_of_ai.pdf (last accessed June 24, 2019).

Nussbaum, B. (2003). African culture and Ubuntu. Perspectives 17: 1-12.

OECD (2017). Digital Economy Outlook 2017: ICT specialists by gender. Available at http://www.oecd.org/internet/oecd-digital-economy-outlook-2017-9789264276284-en.htm (last accessed June 24, 2019).

O'Neil, C. (2016). Weapons of Math Destruction: How Big Data Increases Inequality and Threatens Democracy. New York: Crown Random House.

Ottenberg, S. and Ottenberg, P. (1960). Cultures and Societies of Africa. New York: OUP, pp. 22-35.

Öztürk, P. and Tidemann, A. (2014). A review of case-based reasoning in cognition-action continuum: A step toward bridging symbolic and non-symbolic artificial intelligence. The Knowledge Engineering Review 29(1): 51-77. Available at https://www.cambridge.org/core/journals/knowledge-engineering-review/article/review-of-casebased-reasoning-in-cognitionaction-continuum-a-step-toward-bridging-symbolic-and-nonsymbolic-artificial-intelligence/C5B382C273A5BFF4297A53D96EB5260D (last accessed June 24, 2019).

Pagano, G., Guida, M., Tommasi, F. and Oral, R. (2015). Health effects and toxicity mechanisms of rare earth elements – Knowledge gaps and research prospects. Ecotoxicology and Environmental Safety 115: 40-48.

Pearce, L.H. (2016). Burning Woman. Cork, Ireland: Womencraft Publishing.

Podobnik, B., Kirbis, I.S., Koprcina, M. and Stanley, H.E. (2019). Emergence of the unified right- and left-wing populism—When radical societal changes become more important than ideology. Physica A: Statistical Mechanics and its Applications 517: 459-474.

Potter, C. (2014). How to Make a Human Being: A Body of Evidence. London: Fourth Estate.

Prahbu, R. (2015). Big data – Big trouble? meanderings in an uncharted ethical landscape. *In*: Fossheim H. and Ingierd H. (eds.), Internet Research Ethics. Hellerup: Cappelen Damm Akademisk, pp. 157-172.

Presidente, G. (2017). Labor Services at Will#8232; Regulation of Dismissal and Investment in Industrial Robots. Paper presented at NBER Summer Institute, 15 July.

Querejazu, A. (2016). Encountering the pluriverse: Looking for alternatives in other worlds. Revista Brasileira de Política Internacional 59(2). Available at http://www.scielo.br/scielo.php?pid=S0034-73292016000200206&script=sci_arttext (last accessed June 24, 2019).

Quinn, D. (1992). Ishmael: An Adventure of Mind and Spirit. New York: Bantam.

Rae, J. (1895). Life of Adam Smith. London: Macmillan and Company.

Rajalakshmi, R., Subashini, R., Anjana, R.M. and Mohan, V. (2018). Automated diabetic retinopathy detection in smartphone-based fundus photography using artificial intelligence. Eye 32(6): 1138.

Revette, A.C. (2017). This time it's different: Lithium extraction, cultural politics and development in Bolivia. Third World Quarterly 38: 149-168.

Reyes, J.W. (2007). Environmental policy as social policy? The impact of childhood lead exposure on crime. The B.E. Journal of Economic Analysis and Policy 7(1). Article 51. Available at https://www.nber.org/papers/w13097.pdf (last accessed June 24, 2019).

Reynolds, M. (2018). Biased policing is made worse by errors in pre-crime algorithms. New Scientist 27th April, 2018. Available at https://www.newscientist.com/article/mg23631464-300-biased-policing-is-made-worse-by-errors-in-pre-crime-algorithms/ (last accessed June 24, 2019).

Richardson, K. (2015). An Anthropology of Robots and AI: Annihilation, Anxiety and machines. Abingdon: Routledge.

Rigby, M., Park, S., Saito, T., Western, L.M., Redington, A.L., Fang, X., Henne, S., Manning, A.J., Prinn, R.G., Dutton, G.S., Fraser, P.J., Ganesan, A.L., Hall, B.D., Harth, C.M., Kim, J., Kim, K.-R., Krummel, P.B., Lee, T., Li, S., Liang, Q., Lunt, M.F., Montzka, S.A., Mühle, J., O'Doherty, S., Park, M.-K., Reimann, S., Salameh, P.K., Simmonds, P., Tunnicliffe, R.L., Weiss, R.F., Yokouchi, Y. and Young, D. (2019). Increase in CFC-11 emissions from Eastern China based on atmospheric observations. Nature 569(7757): 546-550.

Robbins L. (1935). An Essay on the Nature and Significance of Economic Science. London: Macmillan and Co.

Rogers, C.R. (1959). A Theory of Therapy, Personality, and Interpersonal Relationships: As Developed in the Client-centered Framework. New York: McGraw-Hill.

Rogers, G.B., Keating, D.J., Young, R.L., Wong, M.L., Licinio, J. and Wesselingh, S. (2016). From gut dysbiosis to altered brain function and mental illness: Mechanisms and pathways. Molecular Psychiatry 21(6): 738.

Rothschild, M., Rowan, M.R. and Fairbairn, J.W. (1977). Storage of cannabinoids by *Arctiacaja* and *Zonocerus elegans* fed on chemically distinct strains of *Cannabis sativa*. Nature 266: 650-651.

Sackett, R.D.F. (1996). Time, Energy, and the Indolent Savage: A Quantitative Cross-Cultural Test of the Primitive Affluence Hypothesis. PhD dissertation, Department of Anthropology, University of California, Los Angeles.

Sagan, C. (1997). Billions and Billions: Thoughts on Life and Death at the Brink of the Millennium. New York: Ballantine Books.

Saith, A. (2007). Goals set for the poor, goalposts set by the rich. International Institute of Asian Studies Newsletter 45: 12-13.

Sakr, S. (2016). On the cutting edge: Artificial intelligence in medicine. Available at https://www.mcgill.ca/library/files/library/sakr_surya_2016.pdf (last accessed June 24, 2019).

Salmon, F. (2012). The formula that killed Wall Street. Significance 9(1): 16-20.

Salovey, P. and Mayer, J.D. (1990). Emotional intelligence. Imagination, Cognition and Personality 9(3): 185-211.

Sands, J. (1876). Out of the World, or, Life on St Kilda. Edinburgh: MacLachlan and Stewart.

Say, J.B. (1828). Cours Completd' Economie Politique. Paris: Chez Rapilly.

Scheiber, N. (2018). Facebook accused of allowing bias against women in job ads. The New York Times, September 18, 2018. Available at https://www.nytimes.com/2018/09/18/business/economy/facebook-job-ads.html (last accessed June 24, 2019).

Schieber, J. (2019). New study shows human development is destroying the planet at an unprecedented rate. Techcrunch 6th May, 2019. Available at https://techcrunch.com/2019/05/06/new-study-shows-human-development-is-destroying-the-planet-at-an-unprecedented-rate/?utm_source=feedburner&utm_medium=feed&utm_campaign=Feed%3A+Techcrunch+%28TechCrunch%29 (last accessed June 24, 2019).

Schnaiberg, A. (1980). The Environment: From Surplus to Scarcity. New York: Oxford University Press.

Seton, G. (1878). St. Kilda, Past and Present. Edinburgh: W. Blackwood.

Sey, A. and Hafkin, N. (eds). (2019). Taking stock: Data and evidence on gender equality in digital access, skills and leadership. Macau: United Nations University Institute on Computing and Society/International Telecommunications Union.

Shafik, N. and Bandyopadhyay, D. (1992). Economic growth and environmental quality: Time series and cross-country evidence. Background Paper for the World Development Report, 1992. World Bank, Washington DC.

Shoham, Y., Perrault, R., Brynjolfsson, E., Clark, J., Manyika, J., Niebles, J.C., Lyons, T., Etchemendy, J. Grosz, B. and Zoe Bauer, Z. (2018). The AI Index 2018 Annual Report. AI Index Steering Committee, Human-Centered AI Initiative, Stanford University, Stanford, CA, December 2018.

Simon, H. (1965). The Shape of Automation for Men and Management. New York: Harper and Row.

Simon, H.A. (1969). The Sciences of the Artificial. Cambridge MA: MIT Press.

Skene, K.R. (2004). Key differences in photosynthetic characteristics of nine species of intertidal macroalgae are related to their position on the shore. Canadian Journal of Botany 82(2): 177-184.

Skene, K.R. (2009). Shadows on a Cave Wall: A New Theory of Evolution. Forfar, Angus: Ard Macha Press.

Skene, K.R. (2011). Escape from Bubbleworld: Seven Curves to Save the Earth. Forfar, Angus: Ard Macha Press.

Skene, K.R. (2018). Circles, spirals, pyramids and cubes: Why the circular economy cannot work. Sustainability Science 13(2): 479-492.

Skene, K. and Murray, A. (2017). Sustainable Economics: Context, Challenges and Opportunities for the 21st-century Practitioner. Oxford: Routledge.

Skene, K. and Malcolm, J. (2019). Using the SDGs to nurture connectivity and promote change. The Design Journal 22(sup 1): 1629-1646.

Slowan, A. and Croucher, M. (1981). Why robots will have emotions. *In*: Drinan, A. (ed.), Proceedings of the Seventh International Joint Conference on Artificial Intelligence. Volume I. Vancouver: University of British Columbia, pp. 197-202. Available at https://www.ijcai.org/Proceedings/81-1/Papers/039.pdf (last accessed June 24, 2019).

Smith, A. (1759). The Theory of Moral Sentiments. London: A. Millar, in the Strand.

Smith, A. (1776). An Inquiry into the Nature and Causes of the Wealth of Nations, in Two Volumes. London: W. Strahan and T. Cadell, in the Strand.

Smith, A. (1904, originally 1776). An Inquiry into the Nature and Causes of the Wealth of Nations. Methuen and Co. Ltd., Book IV, Chapter 2, Paragraph IV.

Smolensky, P. (1988). On the proper treatment of connectionism. Behavioral and Brain Sciences 11(1): 1-74.

Solon, O. (2016). Team of hackers take remote control of Tesla Model S from 12 miles away. Guardian, 20th September, 2016: Available at https://www.theguardian.com/technology/2016/sep/20/tesla-model-s-chinese-hack-remote-control-brakes (last accessed June 24, 2019).

Solzhenitsyn, A.I. (1968). The Cancer Ward. New York: Dial Press.

Southern Medical Association (2017). Artificial intelligence in health care – Audio/visual biomarkers aid in diagnosis. 18th January, 2017.

Sparrow, R. and Sparrow, L. (2006). In the hands of machines? The future of aged care. Minds and Machines 16(2): 141-161.

Stapledon, W.O. (1930). Last and First Men. London: Methuen and Company Limited. Available at https://upload.wikimedia.org/wikipedia/commons/5/5c/Last_and_First_Men.pdf (last accessed June 24, 2019).

Steel, T. (1975). The Life and Death of St Kilda. Glasgow: Fontana.

Steinbeck, J. (1939). The Grapes of Wrath. Available at https://nisbah.com/summer_reading/grapes_of_wrath_john_steinbeck2.pdf (last accessed June 24, 2019).

Stern, D.I. (2003). International Society for Ecological Economics Internet Encyclopedia of Ecological Economics: The Environmental Kuznets Curve. Department of Economics, Rensselaer Polytechnic Institute.

Stern, N. (2009). A Blueprint for a Safer Planet: How to Manage Climate Change and Create a New Era of Progress and Prosperity. London: Bodley Head.

Sternberg, R.J. (1997). Successful Intelligence: How Practical and Creative Intelligence Determine Success in Life. New York: Plume.

Stewart, W. (1942). Collision Orbit. Astounding Science Fiction Magazine, July, 1942. New York: Street and Smith Publications Inc., pp. 80-107.

Stirling, A. (2007). Risk, precaution and science: Towards a more constructive policy debate: Talking point on the precautionary principle. EMBO Reports 8(4): 309-315. Available at https://onlinelibrary.wiley.com/doi/pdf/10.1038/sj.embor.7400953 (last accessed June 24, 2019).

Sudo, N., Chida, Y., Aiba, Y., Sonoda, J., Oyama, N., Yu, X.N., Kubo, C. and Koga, Y. (2004). Postnatal microbial colonization programs the hypothalamic-pituitary adrenal system for stress response in mice. The Journal of Physiology 558: 263-275.

Sumpter, D.J.T. and Pratt, S.C. (2009). Quorum responses and consensus decision making. Philosophical Transactions of the Royal Society of London B: Biological Sciences 364: 743-753.

Surowiecki, J. (2004). The Wisdom of Crowds. London: Little, Brown.

Sweatt, J.D. (2016). Chromatin controls behaviour: Dynamic regulation of chromatin remodelling controls learning and memory. Science 353: 218-219.

Tagkopoulos, I., Liu, Y. C. and Tavazoie, S. (2008). Predictive behavior within microbial genetic networks. Science 320: 1313-1317.

Tannock, G.W. (1995). Normal Microflora: An Introduction to Microbes Inhabiting the Human Body. London, Chapman and Hall.

Taylor, F. W. (1911). Principles of Scientific Management. New York: Harper and Brothers.

Thomas, R.K. (1998). Lloyd Morgan's canon. *In*: Greenberg, G. and Haraway, M.M. (eds.), Comparative Psychology: A Handbook. London: Garland Publishing Inc., pp. 156-163.

Thorndike, E.L. (1920). Intelligence and its uses. Harper's Magazine 140: 227-235.

Tolstoy, L. (1900). Three methods of reform: Socialism, anarchy, Henry Georgism and the land question, Communism, etc. *In*: *Pamphlets*. Christchurch, Hants: The Free Age Press, p. 29.

Trewavas, A. (2017). The foundations of plant intelligence. Interface Focus 7(3) 20160098. Available at https://royalsocietypublishing.org/doi/pdf/10.1098/rsfs.2016.0098 (last accessed June 24, 2019).

Troeh, F.R., Hobbs, J.A. and Donahue, R.L. (2004). Soil and Water Conservation for Productivity and Environmental Protection. Fourth Edition. Upper Saddle River, NJ: Prentice Hall.

Tronto, J. (1993). Moral Boundaries: A Political Argument for an Ethic of Care. New York: Routledge.

Turchin, A. (2015). A map: AGI failures modes and levels. Less Wrong. July 10 2015. Available at http://lesswrong.com/lw/mgf/a_map_agi_failures_modes_and_levels/ (last accessed June 24, 2019).
Turing, A.M. (1950). Computing machinery and intelligence. Mind 59(236): 433-460.
UN (1982). World Charter for Nature. United Nations General Assembly Resolution 37.7. Available at http://www.un.org/documents/ga/res/37/a37r007.htm (last accessed June 24, 2019).
UNOOSA (2019). United Nations Register of Objects Launched into Outer Space. Available at http://www.unoosa.org/oosa/en/spaceobjectregister/index.html (last accessed June 24, 2019).
USNRC (2010). Advancing the Science of Climate Change. Washington DC: National Academies Press.
U.S. Department of Labor, Bureau of Labor Statistics (2016). Employed persons by detailed occupation, sex, race, and Hispanic or Latino ethnicity. Current Population Survey, 2015 Edition. Available at http://www.bls.gov/cps/tables.htm#annual (last accessed June 24, 2019).
VoPham, T., Hart, J.E., Laden, F. and Chiang, Y.Y. (2018). Emerging trends in geospatial artificial intelligence (geo AI): Potential applications for environmental epidemiology. Environmental Health 17(1): 40. Available at https://ehjournal.biomedcentral.com/articles/10.1186/s12940-018-0386-x (last accessed June 24, 2019).
Waage, J., Banerji, R., Campbell, O., Chirwa, E., Collender, G., Dieltiens, V. and Unterhalter, E. (2010). The millennium development goals: A cross-sectoral analysis and principles for goal setting after 2015. Lancet 376: 991-1023.
Wagner, R. (1880). Religion and Art, Prose Works. Volume V. Lincoln: University of Nebraska Press.
Walker, D. (2002). Real Localism. London: The Smith Institute.
Wanger, T.C. (2011). The lithium future—resources, recycling, and the environment. Conservation Letters 4(3): 202-206.
Waters, R. (2018). Google withdraws from $10bn Pentagon JEDI contract contest. Financial Times, 8th October, 2018. Available at https://www.ft.com/content/5d680566-cb46-11e8-b276-b9069bde0956 (last accessed June 24, 2019).
Wechsler, D. (1958). The Measurement and Appraisal of Adult Intelligence. Fourth Edition. Baltimore: Williams and Wilkins.
WEF (2018). The Global Gender Gap Report 2018. Geneva: World Economic Forum. Available at http://www3.weforum.org/docs/WEF_GGGR_2018.pdf (last accessed June 24, 2019).
Weir, W. (2009). History's Greatest Lies. Beverly, MA: Fair Winds Press, pp. 28-41.
Weise, E. (2015). Computer expert hacked into plane and made it briefly fly sideways, according to FBI. Independent, 17th May, 2015. Available at https://www.independent.co.uk/news/world/americas/computer-expert-hacks-into-plane-and-makes-it-fly-sideways-according-to-fbi-10256145.html (last accessed June 24, 2019).
Weiser, M. (1991). The computer for the twenty-first century. Scientific American September, 1991: 94-104.
Welter, V. M. (2002). Biopolis—Patrick Geddes and the City of Life. Boston, MA: The MIT Press.
West, A., Clifford, J. and Atkinson, D. (2018). "Alexa, build me a brand". An investigation into the impact of Artificial Intelligence on branding. The Business

and Management Review 9(3): 321-330. Available at http://www.abrmr.com/myfile/best_track/conference_89787.pdf (last accessed June 24, 2019).

Westerhoff, H.V., Brooks, A.N., Simeonidis, E., García-Contreras, R., He, F., Boogerd, F.C., Jackson, V.J., Goncharuk, V. and Kolodkin, A. (2014). Macromolecular networks and intelligence in microorganisms. Frontiers in Microbiology 5, Article 379. Available at https://www.frontiersin.org/articles/10.3389/fmicb.2014.00379/full (last accessed June 24, 2019).

Whitaker, R.M., Colombo, G.B. and Rand, D.G. (2018). Indirect reciprocity and the evolution of prejudicial groups. Scientific Reports 8(1): 13247. Available at https://www.nature.com/articles/s41598-018-31363-z (last accessed June 24, 2019).

Wiener, N. (1960). Some moral and technical consequences of automation. Science 131 (3410): 1355-1358.

Wilks, T. (2005). Social work and narrative ethics. British Journal of Social Work 35: 1249-1264. Available at https://doi.org/10.1093/bjsw/bch242 (last accessed June 24, 2019).

Williams, G.C. (1966). Adaptation and Natural Selection: A Critique of Some Current Evolutionary Thought. Princeton: Princeton University Press.

Willshaw, D. (1994). Non-symbolic approaches to artificial intelligence and the mind. Philosophical Transactions of the Royal Society of London. Series A: Physical and Engineering Sciences 349(1689): 87-102.

Wilson, S.W. (1991). The animat path to AI. *In*: Meyer, J.-A. and Wilson, S.W. (eds.), From Animals to Animats: Proceedings of the First International Conference on of Adaptive Behavior. Cambridge, MA: MIT Press.

Wilson, M. (2013). The green economy: The dangerous path of nature commoditization. Consilience 10: 85-98.

Windschuttle, K. (2002). Steinbeck's Myth of the Okies. New Criterion 20(10): 24-32.

Winnie Jr., J. and Creel, S. (2017). The many effects of carnivores on their prey and their implications for trophic cascades, and ecosystem structure and function. Food Webs 12: 88-94.

WISE (2018). Available at https://www.wisecampaign.org.uk/statistics/2018-workforce-statistics/ (last accessed June 24, 2019).

Wojcicki, E. (2019). How to Raise Successful People: Simple Lessons for Radical Results. Boston, MA: Houghton Mifflin Harcourt.

Wolfe, A. (1991). Mind, self, society, and computer: Artificial intelligence and the sociology of mind. American Journal of Sociology 96(5): 1073-1096.

Wong, Z.S., Zhou, J. and Zhang, Q. (2019). Artificial intelligence for infectious disease: Big data analytics. Infection, Disease and Health 24(1): 44-48.

World Bank (2019). World Development Report 2019: The Changing Nature of Work. Washington, DC: World Bank.

Worm, B. (2016). Averting a global fisheries disaster. Proceedings of the National Academy of Sciences 113(18): 4895-4897.

Yamagishi, T., Jin, N. and Kiyonari, T. (1999). Bounded generalized reciprocity: Ingroup boasting and ingroup favoritism. Advances in Group Processes 16: 161-197.

Yampolskiy, R.V. (2012). Leak-proofing singularity – Artificial intelligence confinement problem. Journal of Consciousness Studies 19(1-2): 194-214.

Yampolskiy, R.V. (2013). Artificial intelligence safety engineering: Why machine ethics is a wrong approach. *In*: Müller, V.C. (ed.), Philosophy and Theory of Artificial

Intelligence. Berlin: Springer, pp. 389-396. Available at http://cecs.louisville.edu/ry/AIsafety.pdf (last accessed June 24, 2019).

Yudkowsky, E. (2008). Artificial Intelligence as a positive and negative factor in global risk. *In*: Bostrom, N. and Ćirković, M. (eds.), Global Catastrophic Risks. New York: Oxford University Press, pp. 308-345. Available at https://intelligence.org/files/AIPosNegFactor.pdf (last accessed June 24, 2019).

Zhang, H., Feng, J., Zhu, W., Liu, C., Xu, S., Shao, P., Wu, D., Yang, W. and Gu, J. (2000). Chronic toxicity of rare-earth elements on human beings. Biological Trace Element Research 73(1): 1-7.

Zobel, R.W. (1975). The genetics of root development. *In*: Torrey, J.G. and Clarkson, D.F. (eds.), The Development and Function of Roots. London: Academic Press, pp. 261-275.

Index

U

V

W

Y